Discovering Digital Humanity:
A Practical Guide to Creativity and Innovation in the Digital Age

OR

How I Learned to Stop Worrying and Love Technology

by Tom Haymes

Discovering Digital Humanity:
A Practical Guide to Creativity and Innovation in the Digital Age
OR
How I Learned to Stop Worrying and Love Technology

by Tom Haymes

Copyright © 2022

ISBN — 9781626132047
LCCN — 2020933411

Published by ATBOSH Media ltd.
Cleveland, Ohio, USA

http://www.atbosh.com

For my Father, who always opened doors for me

Table of Contents

Foreword	7
Author's Foreword	13
0.1 Introduction: A Holistic Approach to Technology	19
0.2 What Does it Mean to Be a Digital Human?	35
1.0 Designing Human-Centered Technology Ecosystems	59
1.1 Technology Servants	60
1.2 Turning Digital Time into Human Times	93
1.3 Creating Human-Centered Technology Ecosystems	120
2.0 Technology and Narrative	151
2.1 Unbounded Canvas: Creating Digital Stories	152
2.2 Living in the Panopticon: Consuming Narrative	195
2.3 Gaming the System: Socializing Narrative	234
3.0 Designing Human-Centered Technology Systems	269
3.0.1 Introduction to the Case Studies	270
3.1 The Antifragile Class	272
3.2 The West Houston Institute: Design for Augmentation	293
3.3 Thinking Backward: A Knowledge Network for the Next Century	319
4.0 Designing Homo Digitalis	350
Acknowledgements	369
A Special Afterword on the Pandemic and Digital Humanization	371
End Notes	373
About The Author	387

Foreword

By Dr. Bryan Alexander, Futurist, Senior Scholar at Georgetown University, and author of *Academia Next* and *Universities on Fire*

We are still learning to live and learn in the digital world - and we can do it all much better.

The past few years have clearly demonstrated many ways people can use digital technologies to harm others. Rising surveillance, organized and targeted abuse, surveillance as a business model, cyberbullying, cyberwar, hacking elections, the decline of privacy, platforms happily encouraging the spread of dis- and misinformation - to name just a few problems - present a daunting array of suffering and frustration. They are not signs of humanity learning how to live well online.

At the same time, much of the human race has come to rely on the world of digital networks for a huge range of everyday and positive functions. We communicate globally with friends and strangers online. We meet people online, fall in love, interact, stay in touch, play, and learn together. New art forms appear in the digital world, from computer gaming to podcasting and NFTs. People mobilize online to make changes in the analog world. In fact, we use the digital world to address the digital world's problems: sharing digital and information literacy to combat misinformation, organizing against surveillance, coming up with new ways to protect privacy. Above all, we are learning to create, teach, and learn well with these tools.

Doing well and doing even better with the digital world: this is what the following book is all about. Discovering Digital Humanity is a thoughtful and passionate guide to digital potential. Tom Haymes will take you on expeditions into the past and into the human mind in his roles as teacher, designer, storytelling, and provocateur.

The key to Haymes' approach is his vision of a threefold IdeaSpaces structure within which we can best use digital technology for learning and creativity. He wants us to rethink and redesign our learning spaces, our time for reflection, and our social structures (classes, schools, workplaces) in ways that fully realize the implications of the digital world. Rethinking those three domains is not a new idea, of course, but Haymes offers fresh looks on each, then recombines them in exciting ways.

To make the case for IdeaSpaces, Haymes carefully establishes a historical framework. Part of it rests on the transition between industrial and digital eras right up through the COVID-19 pandemic, teasing out the many ways postindustrial life changes inherited human expectations. The author is unafraid to explore dimensions of economics, psychology, philosophy, politics, and education, all with the grace of a liberal mind and a true storyteller. Haymes adds to that timeline the crucial and criminally underappreciated roles of early internet era visionaries. Engelbart, Nelson, Shannon, Papert, Berners-Lee play a key part in the pages to follow, with their highlight achievements acknowledged yet their deeper work appreciated and commended for its transformative powers.

This book's story is one of liberation. Haymes reminds us of how dehumanizing the industrial era was and remains, and points to the many ways digital technologies can enable and expand human creativity and action. He is no utopian. The text constantly reminds us of the many ways humans have abused each other through the internet, including ways we have structured digital platforms that have powerful bad effects. Yet Haymes never lets go of the many ways we can use digital networks to empower our learning and creativity.

That approach naturally creates a subversive book. Haymes targets old ways of thinking, deeming them industrial ruins or, worse, static castles which ultimately fail in their defensive missions. Instead, the author urges us to embrace the digital world's high speeds, relative ease of creation, flexibility, and ability to undermine hierarchies. His call for liquid social structures alone is nearly anarchistic in its ambition. His demand that we embrace paradigm shifts instead of taking incremental steps is bracing.

This is also a book that is keenly committed to storytelling. Haymes uses narrative to lead us easily into the densest problem thickets, always pointing to the human element of the most abstract or technical ideas. He also draws on multimedia storytelling, introducing thoughtful infographics for each chapter which repay close attention. He turns repeatedly to digital photography as an accessible yet deep wellspring for ideas and examples.

This is where three case studies enter the picture. First, Haymes presents the story of reinventing and re-reinventing his undergraduate government classes. This chapter offers a feast for readers who teach, as it unsparingly describes the many challenges these courses provide for both students and instructor. It tracks the author's process as he walks the talk, integrating IdeaSpaces into improving class. Next, we learn about the West Houston Institute, a bold design for a new learning space. Here we see the IdeaSpaces emphasis on, well, space. This is a rich chapter for anyone rethinking or designing a learning space. The third case study looks at how the visions of Bush, Engelbart, and Nelson could be used to develop a new way of creating knowledge, the underpinning of our technological society, using digital tools instead of relying on our current mechanisms of conferring legitimacy to information.

For all of my insistence on the book's engagement with the hands-on work of teaching, not to mention a brew of human history, technology, and psychology, Discovering Digital Humanity is at the same time a deeply personal book. Tom Haymes is our guide and narrator, but also a character whose digital experiences ground some of the discussions. At times, the book reads like an autobiography of someone who has spent decades working at the intersection of education and technology, giving us a very personal grounding for the full sweep of time. Haymes, the college faculty member, shines forth as a passionate teacher, always experimenting with ways of improving students' voyages through his classes. Haymes, the stellar photographer, is another side of his character, offering unique visions of the techno-pedagogical and creative landscapes. His willingness to share his story adds even more humanity to the tale, along with humor and charm. His restless imagination and ability to change his own practices are comforting and inspiring.

We are still learning to live and learn in the digital world, with all of its dangers and triumphs. We do need to rethink our use of space, time, and social structures, and IdeaSpaces is a rich framework with which to do that. Tom Haymes is an excellent guide to this transformation. Onward!

Bryan Alexander
Manassas, VA, October 2021

Author's Foreword

By most measures, I started my journey into a digital existence in late 1981. My father bought an Apple][+ and I began to explore the digital world. In the first year of my use of the Apple computer, however, an earlier computer experience drove me to explore my new tool. In 1977, when I was 10, I went to a party with my parents at someone's house who worked for Rice University. They had a terminal hooked up to the Rice mainframe. In order to distract a 10-year-old, they let me play a Star Trek game that consisted primarily of entering a sector of space, seeing if there were any Klingons, and then engaging in battle with them. As part of this game, you had to monitor the operations of the *Enterprise* and issue commands to your crew. As an avid reader, this was a revelation to me. I could *interact* with a mental story for the first time in my life.

When I got that Apple][, it obsessed me with making it do the same thing as that Rice mainframe had done all of those years earlier. In those early years of computing, software for the Apple (and any computer) was limited and expensive. However, I found a book of software programs and in it was a Star Trek game (to this day I do not know whether it would have had the same functionality as the mainframe version). It was a lot of BASIC code. I spent three days typing it into the computer and debugging all the mistakes I had made, only to have my mother accidentally delete the floppy that I used to store my efforts. Even years later, I searched in vain for an experience similar to what I had experienced on that mainframe computer. I doubt I can ever really duplicate that experience, because my imagination has moved on from that point in my youth.*

There are two things to take away from that story that color my perception of technology. First, I have always viewed technology as a barrier to achieving what I want to achieve. Second, technology is an enabler of my creative drive. I am always seeking to climb (or bypass) that barrier to achieve what I want.

These two elements lead me to the third element of my perception of technology: technology is a means to get me to where I want to go. If I want to go upstairs, the technology of the stairs will get me there. If I want to realize my creative photographic vision, the technology of the camera, lens, film, computer, and print gets me there. If I want to write, the technology of the word processor gets me there. If I want to communicate, the technology of the internet gets me there.

The purpose of this is book to begin a conversation around reshaping our cultural assumptions about technology, its possibilities and opportunities, and how to create personal and organizational structures to make technology work for us, not the other way around. Our global challenges, ranging from climate change to education to democracy, demand that we

---

* My good friend Bryan Alexander found the Wikipedia entry on this game for me. **en.wikipedia.org/wiki/Star_Trek_(1971_video_game)**

create a healthier dialog with our technology and use that to facilitate conversations among humans.

This is not happening. I see instead these days is a set of dysfunctional relationships that have a pernicious effect on ourselves and our ability to grow as a species. Fundamental assumptions about an adversarial relationship with our tools drive narratives around politics, economics, and our very bodies themselves. Instead of reveling in the unprecedented knowledge at our fingertips, we use technology to hide information and manipulate public perceptions. Instead of rejoicing in the possibility of liberating humanity from dangerous and boring tasks, we fixate on the social disruptions of "being replaced by robots." Instead of creating organizations designed to grasp economic and social opportunities, we allow them to act as a drag on the possibilities offered to us by the exponential growth of technological opportunities. This is because we persist in having old conversations under new circumstances. We are not adapting at an organizational, let alone a personal, level.

I do not believe that we will live in the world envisioned by Star Trek any time soon. However, I believe we waste countless opportunities to make ourselves, our families, and our societies better because of the phobias we have about technology. We have done this to ourselves through poor design, breeding false mythologies, and the accretion of power to those who would perpetuate them. I continue to believe that technology, especially information technology, offers us unprecedented possibilities for liberation, both on a personal and societal level. There will be dislocations and political challenges, but we can overcome them with a clear-eyed view of the limitations and opportunities that our technologies provide us. Technology is neither moral nor immoral. It is amoral. It is a canvas upon which we paint. The picture we create depends on us. It's time to pick up the brush.

Tom Haymes
Katy, Texas, June 2021

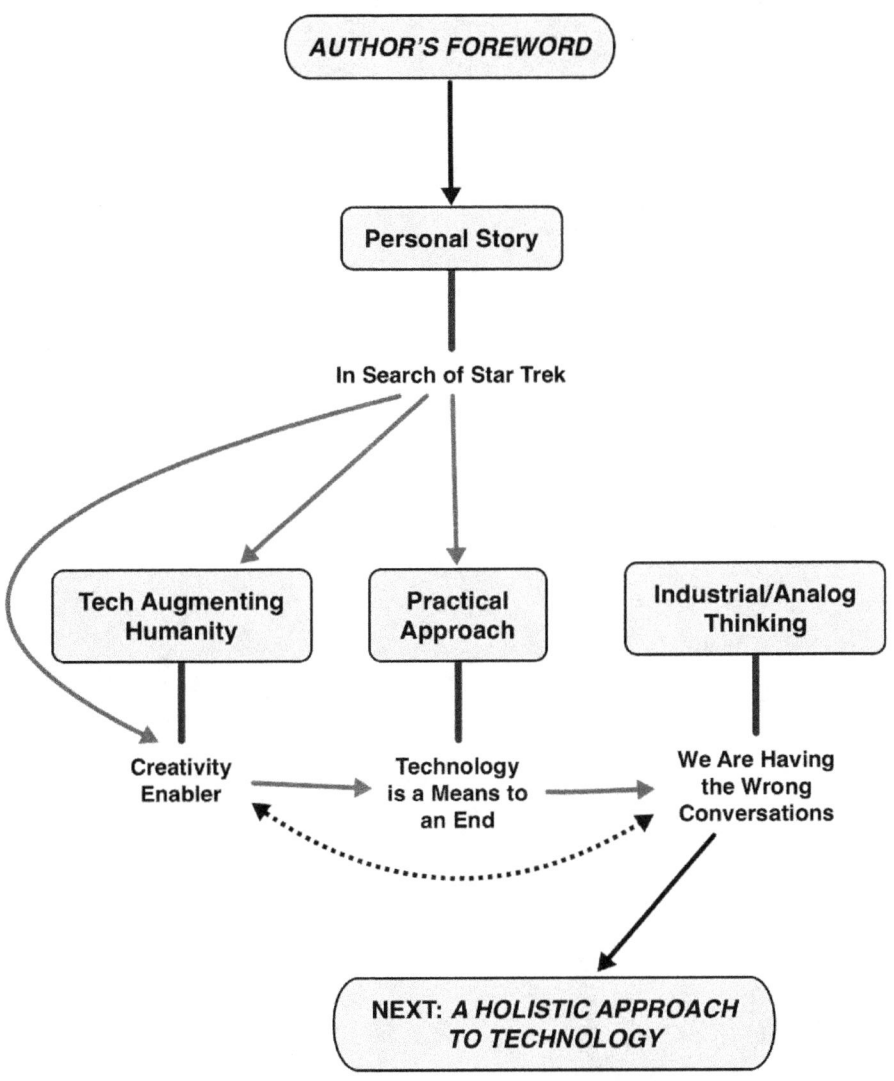

ORGANIZATIONAL NOTE: At the end of each chapter there are one or more concept maps that outline the essential narrative of the chapter and how they relate to the major themes of the book (Industrial vs. Analog Thinking and Technology Augmenting Humanity) as well specific reference to Practical Approaches to Implementing the Ideas of the Book. In the first two sections, the chapters also contain a second map that relates the key themes in the chapter to the IdeaSpaces framework (outlined in the next chapter).

0.1 Introduction:
A Holistic Approach to Technology

For the want of a nail the shoe was lost,
For the want of a shoe the horse was lost,
For the want of a horse the rider was lost,
For the want of a rider the battle was lost,
For the want of a battle the kingdom was lost—
And all for the want of a horseshoe nail.
            Traditional Proverb as Related by Benjamin Franklin

I came into that room with purpose. It was the first day of the semester and I was trying out a new set of teaching strategies. A large part of this lesson included teaching them how to integrate into their processes to augment their learning, both in my class and beyond. The learning space, integrated digitally and physically, would be a workplace where it was possible to design your way out of problems through the creative use of powerful technologies. For this, I wanted them to interact with concept maps online and in person. These tools formed part of a design challenge about real-world problems that the students selected themselves. This was going to be fun. I was raring to go. And then there was no wire to hook up my laptop in the technology system....

This is not an atypical experience for me. I have always pushed the boundaries of our technology systems. What I didn't realize was that this was a much deeper problem than one of simple user interfaces. Benjamin Franklin related the aphorism, "for want of a nail" to explain how a simple oversight in one area can lead to systemic failure (nail —> horse —> rider —> battle —> kingdom, etc.). However, in an age characterized by technological systems vastly more complex than that of a horse and rider (and even this is more complex than the one might assume) we have many, many nails that require our attention.

Let's go back to my challenge. My problems in teaching this class extended well beyond the poorly designed technology system that tried to force me to use the installed PC in the room, which did not have the software I needed to teach the class. That was annoying but fixable because I simply unplugged the PC and used its display cable to interface with my laptop. There were far bigger challenges in my way. At a basic level, the college arranged the room in rows of desks that had to be moved in order to allow for any collaborative group work. Since they were not on wheels, this was possible, but more difficult than necessary. There were no screens for the students to work together online, either. But again, these were fixable problems that degraded the experience but still made it possible to teach this way.

The far bigger problem was that the 90-minute schedule dictated the flow of time in the class. That meant, no matter where the students were in their brainstorming, class was over when the clock ran out. This kind of time boundary led to the temptation to engage in more "traditional" means of instruction, such as lecture, because it was far easier to control the flow of time using that method. People don't learn according to a schedule. Deadlines are fine to focus the mind, but they need to leave enough leeway to explore different pathways to understanding. The traditional semester schedule constrains the natural process of learning. Also, all the time spent rearranging the room took away from true "work" time.

But even this was not the biggest barrier to the success of my students. They were unused to the way I taught, because every other class conformed to outdated norms. Everyone in the larger structure accepted that they were part of the education machine and worked within its confines. Those confines were in themselves an artificial byproduct of the Industrial Age and its conveyor belt view of mass education. While most accept this as being normal, it only goes back a little more than a century. All factories require metrics. This reality led to the far more pernicious side effect on my students: an almost obsessive focus on what grade they were going to get distracted from their actual learning. Again, this is understandable given that the currency of our educational system is that of grades rather than learning. The narrative of the metric wins over human learning almost every time.

Industrial constraints also fail our humanity and hobble our firms as they double down on a fading model and its associated metrics instead of exploring new paradigms of product creation and new metrics of success. We have discovered that the "real" world analogy to grades, profit, can be every bit as pernicious to the long-term success of companies and the externalities they impose on the rest of society (Raworth, 2017).

When I entered the technology world as a child in the 1970s, the limitless possibilities it seemed to offer hooked me. I could write on word processors that didn't care whether I made a mistake. I could (eventually) take photographs that would allow me to see the images in my mind. I could play games that allowed me to create my own stories. Even more earth shaking, I could talk and share with an endless stream of other minds, at first through bulletin boards, but then through an ever-expanding cornucopia of social media.

When I became a teacher, I wanted to use these tools to transform my students into thinkers capable of standing on and with the shoulders of others through the emerging technological tools available to them. I wanted them to learn by making mistakes. I wanted them to innovate their lives as the possibilities that I saw in technology were endless. They still are. The problem is not the technology. It is us.

This challenge extends far beyond education. I have worked in a wide variety of corporate environments, from startups to massive corporations, from technology companies to architecture and public relations firms. In every instance, I have encountered the same barriers that I encountered in that classroom: tasks do no drive the technology, the strictures of time are not as malleable as the technology allows, and evolutionary change is not possible because the various human systems have failed to adapt to the new technological realities.

What we see when we survey the individual and organizational landscape, both inside and outside of education, is example after example of technology being used to reinforce old industrial paradigms. From propaganda infiltrating social media to workers working longer and longer hours under the watchful eye of surveillance technologies to companies that are ever more complex but still exist within specialized industrial silos in their organizational structures.

These are all examples of us wearing industrial blinders in a Digital Age and failing to see the possibilities all around us. Instead of the best of humanity, we have used technology over and over to bring out the worst in us. I think this is in part because of a fundamental disconnect in how our various cultures deal with the rapid pace of change in the world. We need a way of reframing our way of looking at the world. This will require that develop new ways of telling our stories and recast our relationships with our tools. Technology does not exist in a vacuum, but is situated in a set of social constructions of our own making. Approaching tools without context leads to inefficiency or, at worst, perverse outcomes for our humanity. A central argument of **Discovering Digital Humanity** is that we need frameworks to situate technology within the broader context of our human systems.

The IdeaSpaces Framework: An Overview

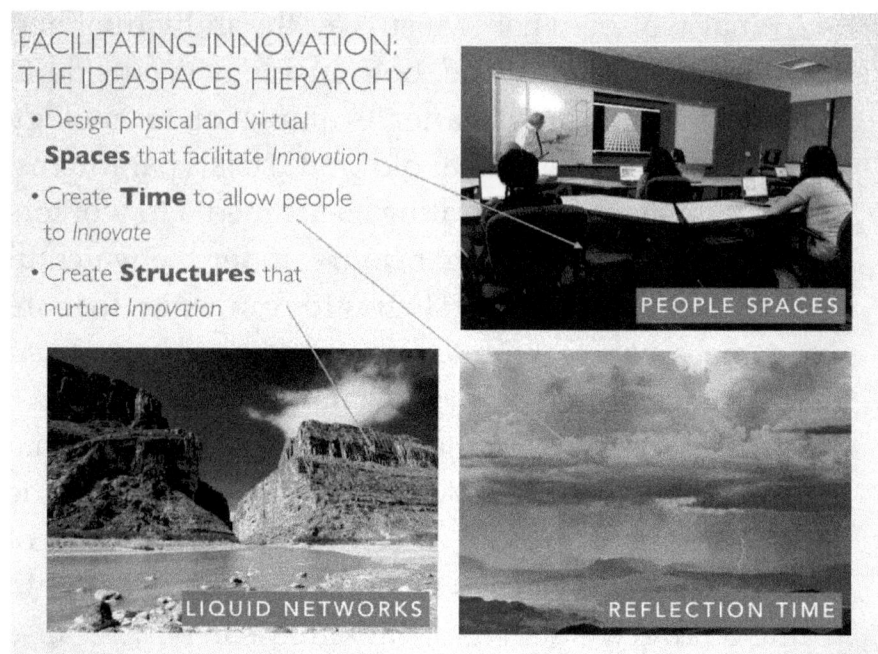

Figure 1 - The IdeaSpaces Framework

IdeaSpaces grows from an observation that I made over the course of my career as a technologist, teacher, social scientist, and administrator: Innovation and change are multi-causal processes that are often missed because of narrow decision-making structures. These are often a legacy of an industrial era that is integrated into our environments at all levels. The IdeaSpaces framework proposes a more holistic approach for understanding the context of learning and innovation in which any technology exists. Its purpose is not to support technology. IdeaSpaces is a framework to analyze and create human environments where good ideas can flourish and grow into lasting change.

There are three levels to the IdeaSpaces framework. At the bottom are physical and virtual ✡ **SPACES** that can range from virtual environments to learning spaces to collaboration spaces to entire buildings. Next up, the level of difficulty is creating ⧗ **TIME**, which is all about creating mental space for employees, students, etc. to get some perspective on what they are doing and to iterate solutions to problems. Finally, and perhaps most difficult, is our ability to reimagine and adapt ❈ **STRUCTURES** to ones that facilitate and nurture Innovation and Change rather than stifle them.

Language is often inadequate to describe the shifts we are seeing in the world. Language is a lagging indicator of historical and paradigmatic shifts as the human imagination needs ⧗ **TIME** to negotiate and assign meanings to the new realities that it must confront. The three IdeaSpaces concepts represent paradigms of thought that we have built up around their respective definitions. Throughout this book when I refer to the larger context surrounding any *space*, *time* or *structure* that are being challenged and reshaped by new digital realities, I will use the symbolism established in this chapter (✡ **SPACE**, ⧗ **TIME**, ❈ **STRUCTURE**) to help the reader understand I am referring to the larger conceptual continuity in which these ideas live, and not referencing their more concrete definitions.

People *spaces* (✪ **SPACE**) apply effective design to our physical and virtual spaces. Effective spaces are a required precondition for innovative environments. However, they are not sufficient. I have been involved in the design or execution of dozens of "innovative" spaces, both physical and virtual, that have failed to live up to their potential because they became the ultimate product of the "innovation" process. The project planners misunderstood technological tools as an end rather than as a means. If "innovation" ends with the constructed environment, it will constrain its potential to ignite the human capacity for learning and growth.

Reflection ⧖ **TIME** is required for the inspiration and iteration of ideas. Eureka moments are the province of fiction. People need ⧖ **TIME** to reflect on what is working and what isn't. Sometimes this occurs, but often group activity is necessary to gain sufficient perspective on a challenge in order to start a meaningful discussion about change. Digital ⧖ **TIME** reduces the cost of iteration.

Because it involves changing lots of interconnected people, ✺ **STRUCTURE** is perhaps the thorniest issue that has to be considered when creating a culture of innovation and change. Even large organizations considered to be innovative, such as Apple or Google, have had to go to great lengths to maintain an edge, as their organizations grew in complexity and size. This challenge is even more difficult for organizations that were formed during the Industrial Age and are already large and bureaucratized, such as institutions of higher education or legacy firms. Organizations are networks of people. Technology improves their capacity for communication. However, it doesn't solve organizational shortcomings. Papering a technology over a broken process doesn't fix that process. It just speeds up that process in the direction it was already traveling.

A second issue here is that the sheer volume of information available to most organizations has gone up exponentially. Steven Johnson studied how innovation occurs and concluded that it is a social process, not an individual one. Some social processes do a better job of distributing information than others. The way an organization reacts to information flows is therefore critical in determining whether it facilitates innovation and learning, or whether it stifles these impulses.

Johnson makes the analogy to physical states to describe three distinct classes of information networks that characterize organizations. *Gaseous Networks* are networks where lots of ideas fly around, but they never achieve meaningful purchase on anything substantive. *Rigidity is* the defining characteristic of Solid Networks. Ideas enter the system but bounce off and nothing changes. Johnson sees the ideal organizational network as a Liquid Network. In this kind of ❋ STRUCTURE, ideas can change the landscape much like a river gouges out a new channel, but there is still sufficient ❋ STRUCTURE to provide purchase for new ideas to take hold. (Johnson 2010, pp. 43-66) *Liquid Network*s describes an organizational

❋ **STRUCTURE** that is flexible enough to allow for change and yet structural enough to capture innovation.

It is only with the mindful application of all three concepts that we can sustain innovation in any organization. I hope that through an ongoing discussion and iteration, the concept of IdeaSpaces itself will help people take a holistic approach to creating the change that will help us and our children survive and thrive in a world of endless possibilities and challenges.

IdeaSpaces is a process of deep reflection and self-examination. Exposing any project to this kind of examination will often reveal unexpected, deep-seated, and complex issues that stand in the way of its success. Organizational issues are often difficult to address. That is why sometimes even the most simple projects require careful analysis of how they fit into your organization's workflow and goals. No technology exists outside of context. New technology will fail if it doesn't address that context. Expecting technology to change your organization is like expecting a nail to fix the battle strategy.

I organized the book through the lens of the IdeaSpaces framework. The first two sections (Digital Technology and Digital Humanity) comprise three chapters that explore the potential to apply ✿ **SPACE**, ⧖ **TIME** and ❋ **STRUCTURE** to create opportunities for our effective adoption of tools and the augmentation of our collective creative potential. The last section comprises three case studies that show

how the three we can incorporate the IdeaSpaces concepts into the designs of spaces, programs, and digital platforms, always putting the human experience in the forefront. I designed the cases as an applied set of roadmaps to rethink the *human element* of our digital existences.

The first section flips our relationship with a wide range of technologies and how they either impede or augment our abilities. It represents a journey along a spectrum that starts with our basic interactions with technology, design considerations, and how we can use technology to augment ourselves. This section provides a framework for demystifying technological interactions and to think about technology as an enabler, not an enslaver. First, we will look at how we can master technology ✷ **SPACE** rather than becoming its slave. Second, we consider the importance of reflection ⧖ **TIME** in how humans might adapt to digital technologies and environments. Finally, we will look at how these kinds of systems can create and foster ecosystems (✾ **STRUCTURE**) where humans and technologies can symbiotically interact to create flourishing innovation and learning.

Stories power our creativity and innovation. We are all storytellers. Storytelling underpins our very humanity as we struggle to make sense of our world. The Digital Age has disrupted the speed and distribution of our stories more than at any time in human history. We're constantly constructing and deconstructing narratives throughout our lives. This is not new and dates back to the dawn of our communication as a species. What has changed is the speed of change. The digital world offers us new tools to communicate faster and across a vastly larger range of media.

In the second section of the book, we will focus on how the Digital Age changes how we tell stories. The interplay of communication and play generates opportunities for innovation and reinvention, but we learn how to turn digital noise into digital art. This is a critical starting point for where we take our technological spaces, as the most restrictive frames we've developed are in our own minds.

The first chapter in the section explores how digital technology explodes our notions of the kinds of narratives we can create (✪ **SPACE**). The second chapter explores the challenges of harnessing our digital networks to create innovation and reinvention instead of misinformation and manipulation (⧖ **TIME**). Finally, we will look at how technology married with narrative creates new play. We use play to explore tough challenges and turn them into opportunities for innovation (✹ **STRUCTURE**). Digital communication provides a key bridge between individuals and organizations for disrupting industrial-era processes.

In the last section of the book, we will come full circle and show how to marry the possible with the doable. The three case studies in this section apply the concepts of the IdeaSpaces framework to practical scenarios.

In the first study, we will look at how I created an antifragile classroom through a holistic design process based on the IdeaSpaces framework that thrived despite the challenges of pandemic remote teaching. I did this by integrating a holistic analysis of the tools I was using (✿ SPACES), both physical and virtual. I complemented this by carefully analyzing how I used ⧗ TIME in my instruction and adapted it to the requirements of interacting with my students via videoconferencing. Finally, I created ✸ STRUCTURES designed to support my students' motivation as well as provide a digital roadmap to keep them on track during a physically isolated learning process.

The second case study will examine how the IdeaSpaces framework can apply to the design of a physical space. The West Houston Institute was a purpose-built ✿ SPACE designed to foster intersectional behavior. We coupled this with programs that created ⧗ TIME for users of the building to innovate and reflect. We also placed the building into an integrated ✸ STRUCTURE to create structural incentives for ideas and people to collide, learn, and innovate.

The final case study will explore how we could design a virtual network of ideas within the IdeaSpaces framework. The Deep Thought project is an aspirational design for creating an online ✿ SPACE based on concepts developed by Vannevar Bush, Douglas Engelbart, and Ted Nelson for information exchange, which gives its users ⧗ TIME to exchange and evaluate information comprehensively and transparently. I explicitly designed it to create a new ❈ **STRUCTURE** of knowledge exchange based on digital concepts rather than the present analog ones.

First, however, let us turn to what it means to be human in a digital age.

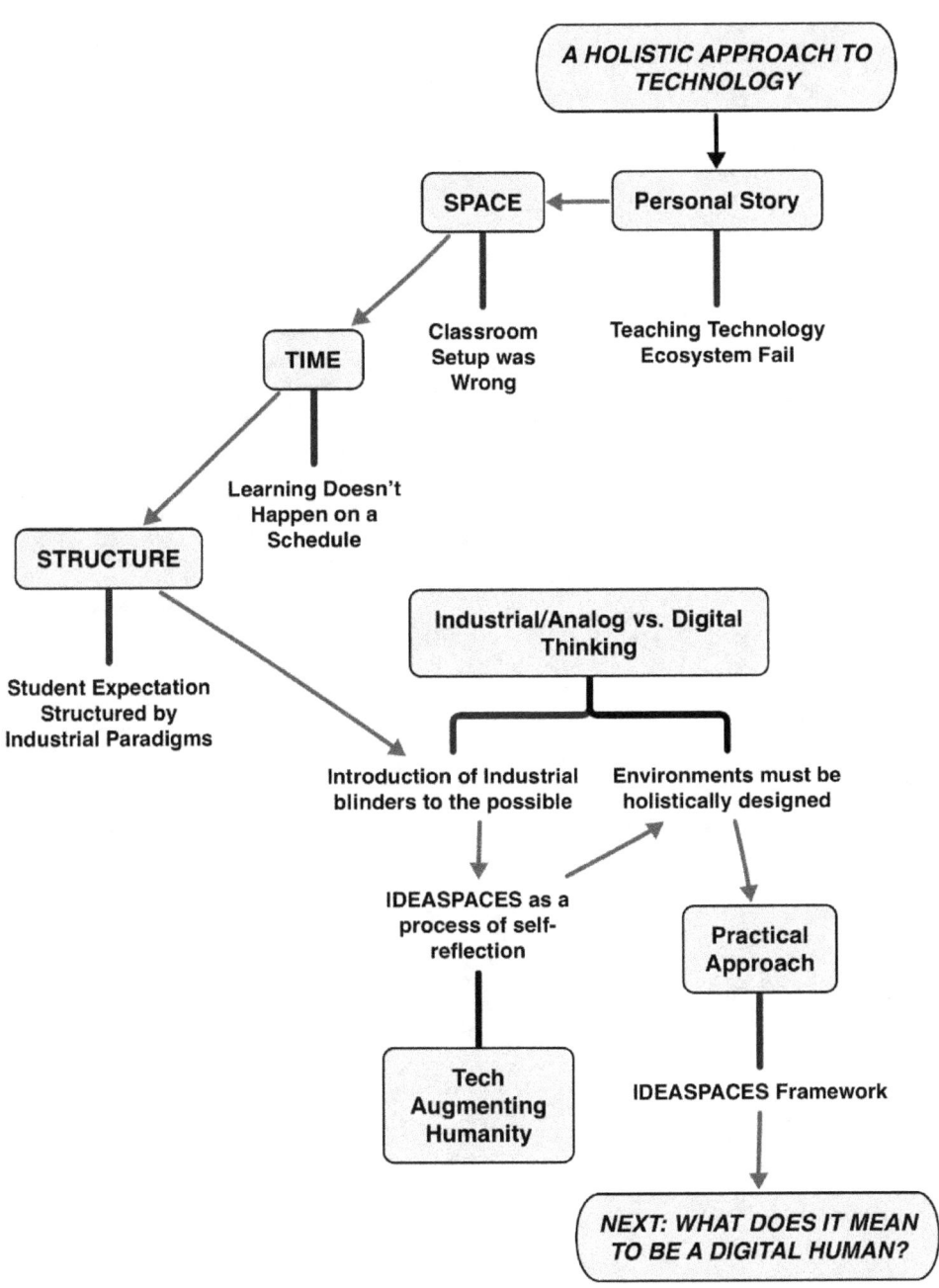

0.2 What Does it Mean to Be a Digital Human?

"Did you win your sword fight?"
"Of course I won the fucking sword fight," Hiro says.
"I'm the greatest sword fighter in the world."
"And you wrote the software."
"Yeah. That, too," Hiro says.

Neal Stephenson, 1993

It's hard to know when I transitioned to thinking digitally. At its root, digital thinking is innate. We all start as digital thinkers. It is a belief that the world is malleable to change and that you can adapt it to fit your needs. At the scale of children, our worlds are small and infinitely malleable. At some point, these toys become more structured and start to undermine our belief in constructed reality. Some toys are better at sustaining this constructive attitude than others. As a child I played with Legos (the basic sets, not the "kits") and as a teenager I became fascinated with Dungeons and Dragons. Both activities rely on an openness to constructed realities.

The computer simply extended my ability to construct a set of tools that enabled me to create everything from stories to photographs to maps of ideas. My *homo digitalis* self is therefore nothing more than an extension of the world building I did as a child. The digital world allowed me to never grow up and to continue playing. My playground just expanded from the playroom floor to the word processor to the internet. I have always believed that I could shape the world to my needs. This has not left me.

We are at our most human as a child. School then conditions us to conform to society. If that society is industrial, it teaches us to conform to the machines that we will eventually work in. Those systems have been relatively fixed for over a century now. What goes through an office or factory may have changed, but our relationship to the office or shop floor has not significantly changed for a century or more. This is a lesson I resisted learning because, to me, the machine was always my servant. Like my Legos, I could use it to create new realities.

I am a creator of worlds. This is what makes me digital. These are worlds of creativity and imagination; worlds of new constellations of ideas; worlds of fresh stories to explore and follow. Computers gave me the metaphor of a software world and they gave me the tools to express my ideas and those of others in new ways. They did not make me digital. The child was always there.

We are already *homo digitalis*. It's just locked away inside the years of indoctrination we've had becoming *homo industrialis*. Learning to become digital is relearning how to become human. In order to find our way back, we must learn to use machines to amplify human stories, not dictate them.

Machines told the story of the Industrial Age. It was often art that showed us this most clearly. In 1936 Charlie Chaplin made "Modern Times," the last of his great silent films. In it, the machinery of the Industrial Age systematically consumed the Little Tramp character. He is turned into a human bolt tightening machine, force fed by a robotic torture device, and eventually dropped into the machinery of the factory itself. Chaplin's creative genius neatly expresses the dehumanizing nature of an age where humans were subsumed to the whims of the larger machine, whether that be a factory or a bureaucracy. What has happened since then is that many of those dehumanized mechanical jobs have been automated and humans have largely been removed from the manufacturing floor. The Industrial Age isolated us from our tools and made us into gears, not gods.

Less than a quarter century later, however, we saw the beginnings of an automation of a different kind. Instead of our arms and legs, it was now our brains that were becoming part of a larger machine. This was even more dehumanizing than the world that seemed to consume the Little Tramp. Replacing the gears and levers of Chaplin's factory was a new form of assembly line. We built this one from vacuum tubes.

Imagine a world where your interaction with the digital environment was mediated by a group of priests whose job was to take your ideas, feed them into a computer, and return, often days later, with, if you're lucky, a useful answer. If you weren't lucky, they would simply return with an error message. In the first 20-30 years of the Digital Age, this is what computers resembled. We processed ideas just like Chevys. Worse yet, most of us had a poor understanding of what was happening in those black boxes. Even science fiction authors like Isaac Asimov and Fredric Brown considered large, unwieldy computers as the norm. They saw the power of computers in the same context as Henry Ford viewed his car plants: bigger is better. This view of computers was both intimidating and distant. We did not see them as enablers of human ingenuity, like a hammer and chisel in the hands of a skilled carpenter. Instead of familiar, empowering tools, we saw computers as threatening and dehumanizing. Brown envisioned such a scenario in his short story "Answer." All the computers on the planet were tied together and asked, "Is there a God?" The answer was "Yes, *now* there is a God." (Brown 1954, p. 23) Thankfully, this is not the world we live in today (or is it?).

Thanks to several visionaries that saw past this industrial version of computing technology, we have made a lot of progress toward lowering the barriers between ourselves and our ability to create new worlds. One such figure was a frustrated engineer named Douglas Engelbart. In 1962, he was creating a proposal for a new research effort. Engelbart understood that the computing technology of batch processing, punch cards and mainframes further distanced humans from their creative efforts. He saw in the digital world an opportunity to break us free of the shackles imposed by the industrial one. Engelbart and others such as J. C. R. Licklider, Ivan Sutherland, and Bob Taylor saw beyond the "big computer" paradigm and realized that this way of thinking about computing technology was a fundamental misapplication of its possibilities. Engelbart's proposal focused on breaking down the gulf between user and tool.

Engelbart called his project "Augmenting Human Intellect" and he proposed a research effort specifically targeted at human-centric technologies for computer interface. His subsequent effort resulted in the creation of the mouse, practical hyperlinks, a graphical user interface, videoconferencing, and many other innovations, all within five years. His "Mother of All Demos" in 1968 showed what was possible when we designed computing around the human instead of an industrial paradigm that dictated that humans must conform to the needs of the machine. He charted a pathway to a truly "personal" computer, which we now see as the norm. Several of the engineers from his Stanford Research Institute (SRI) moved on to Xerox's Palo Alto Research Center (PARC) in the 1970s and there created the Alto. The Alto led to the Macintosh of the 1980s and the widespread dissemination of graphical user interfaces in the 90s.

Engelbart's larger goal still eludes many of us almost 60 years on. Engelbart thought until his death that the possibilities of augmenting human intellect through the application of digital technology remained unfulfilled. He said in 1999 that, "We need to stimulate the active coevolution of the human system and the tool system.... The goal of augmenting rather than automating is being able to stimulate and design this coevolution.... That would let me retire, if I could see a really effective pursuit of that collective IQ-raising." (Rheingold 2000, pp. 327, 329) While he solved many of the initial problems of ✪ **SPACE**, Engelbart's project of augmenting human intellect has floundered on the human part of that equation. By making computing machinery more approachable, Engelbart's innovations enabled massive productivity growth but, ironically, did not undermine the cultural norms of the very ❀ **STRUCTURES** that Engelbart saw holding humans back.

Doug Engelbart understood that the technical problems, such as they are, stemmed from a mismatch between technology and intentions. Throughout his prolific period of invention in the 1960s, the human goal of augmenting our intellects was the one thing that drove him. He viewed this as a cooperative project between man and machines. All of his other innovations were secondary to this goal and were measured against it.

It was when Engelbart started taking on ❀ **STRUCTURES** after he left the Stanford Research Institute (SRI) that he realized that his technological efforts had only scratched the surface of the problem. Organizations can stifle just about any technological achievement to augment our humanity. Engelbart's team did much to achieve his vision through their own heavy lifting at SRI in the 60s. However, once they had established the groundwork of basic interface design (the mouse, hyperlinks, and GUI foremost among them), the ability to exploit these machines to augment humanity passed to the rest of us. We allowed ourselves to become enslaved to productivity instead of using these technologies to create new human potential. We merely made human beings into more efficient widgets using the tools that Engelbart and others provided instead of using them to rediscover our humanity in the widgets we had become during the Industrial Age. As Howard Rheingold, Engelbart's interlocutor in the aforementioned 1999 interview, wrote elsewhere in that same volume:

> Even if sufficient resources could be allocated to making the best educational use of the Net, the problem remains of teaching tomorrow's citizens how to think for themselves. Parents aren't keen on teaching their children a critical and questioning attitude, and teachers aren't equipped to show children how to find their own way through the new world where you don't trust the publisher of information to tell the truth. (Rheingold, 2000, p. 322)

These are self-inflicted ❋ STRUCTURAL constraints. Technology cannot help us if we insist on limiting our vision of what is possible (or desirable). While trying to give us the tools to escape the slavery of the assembly line, Engelbart only gave us digital wrenches that allowed us to more efficiently become unquestioning parts of the thinking machines. Instead of being our partners, they remained our masters. Chaplin's dystopian vision of the Industrial paradigm still enslaves us. It has just gained digital trappings.

We still measure technology through productivity instead of the human growth it provides. For much of the Industrial Age, productivity meant that humans gained new freedoms. For instance, the productivity of home technology made it possible for women to more positively contribute to society. However, since the 1970s, productivity has been used to exact more and more work from each worker, women included. Now, the productivity of the machines threatens to eliminate the human element altogether. This is very dangerous because the ability of these new tools to manipulate, mislead, and dehumanize is unprecedented.

Never in human history has the possible outstripped the doable to such a degree. The key thing that has changed since Chaplin's time is that all technology, driven by the last wave of data technologies, is now almost completely fungible. Unlike past revolutions, which involved physical things from factories to weapons to cars, this one is a revolution of data and code. We often compare the information convulsions of the Digital Age to the print revolution of the 15th and 16th centuries, but limited literacy and the physical nature of print media acted as brakes on the dissemination of books, pamphlets, and newspapers. Now, we have Russia using digital technology developed in the United States to attack the United States within a matter of years rather than decades. Arguably, it's a matter of weeks and months on its smallest scale. Our industrial media and political systems cannot keep up.

Even the Russian hack of the 2016 election is still "old" technology by today's standards. Most of the tools used by the Russians were evolutions of technologies developed in the last decade, most notably algorithms designed to facilitate the sale of what is essentially advertising (an even older technology) through media such as Facebook, Google, Amazon, and Twitter. The failures of the 2016 US election cycle, as well as similar failures worldwide, were profoundly human. The Russians simply executed old KGB desinformatsia strategies developed during the Cold War. Digital tools amplified their effect. The speed and power of our tools reshape old strategies and give them new capabilities unimaginable to their original creators.

Our business, education, and government organizations simply cannot cope with the pace of change and are blindsided by relatively simple tactics. Again, these represent human and organizational failures to look past the technology and understand what it's truly doing.

The Russian case is an example of something that will be somewhat of a sub-theme to this volume. Most of our "technological" failures are actually the product of the application of new, faster, more lethal technologies on top of older, industrial-era paradigms. Digital tools can be used to liberate workers and create new opportunities for all. They can also be used to create surveillance, misinformation, and propaganda mechanisms that may finally realize an Orwellian nightmare.

The best hope we have individually and collectively to realize a better world for ourselves and our children is to figure out how to disconnect these tools from industrial, analog paradigms and connect them up to new digital paradigms of transparency, recombination, and augmentation. Either way, the pace of technological change will continue to move forward relentlessly. How we respond to it is key to the world we, and our children, end up creating.

Technological momentum increases as it becomes more lightweight. Technology changes ✲ **SPACE**. Technology changes ⧗ **TIME**. Technology doesn't change our mental and organizational ❋ **STRUCTURES**. We must do that ourselves. This equation has only changed in speed because edge cases can access and manipulate technology to disrupt complex processes. However, the reactive ❋ **STRUCTURES** we exist in have not been able to adapt to new realities. You don't need to retool a factory to develop an algorithm. You don't need to change a supply chain to reconfigure a database of user information. Factories are a lot easier to regulate than algorithms.

No matter how much teeth gnashing may go on in legislative bodies around the world, we cannot slow the pace of change. You may retard it temporarily in your country, but there is very little to keep it from relocating to a more amenable venue. If Congress were to decide to regulate Facebook, there's little to stop it from moving its operations to the Philippines. Even if you could block it, the underlying drives that

Facebook currently meets could be easily picked up by a competitor, rebranded, and relaunched as a new service with blinding speed. Snapchat is replaced by Instagram, which is in turn replaced by TikTok. The genie is out of the bottle. The limiting factor is the human network, not the technology itself as in the Industrial Age.

Also, these fundamental technologies are moving to other areas we don't traditionally associate with computing. We already have widespread, ubiquitous computing with the migration of significant computing power to mobile devices. Those smartphones are now miniaturizing even more into Internet-of-Things (IoT) devices that are further atomizing the invisible web we are constructing around us. Meanwhile, at the other end of the spectrum, more powerful centralized "brains" can process ever more complex problems (increasingly receiving data from the aforementioned IoT devices) to attack an expanding range of brute force issues from climatology to oncology to sociology. These realities are alternately thrilling, empowering, and profoundly disturbing. This is even more true as data technology moves into the most profoundly human areas of all, even down to our very genetic code.

Our ability to unleash tools of understanding into more and more areas has created many digital opportunities. However, our mindsets continue to hamper our ability to adapt to the new realities that are being created daily. In order to become more digital, we must actively reimagine the systems that we take for granted. Consider the many ways in which we are hostage to relics of the Industrial Age, from standardized time zones (and Daylight Savings Time) to consumerism to the business cycle itself. Many (but not all) of the limits we have encountered have been products of inflexible mindsets more than inflexible technologies. It's amazing how often we developed creative solutions using available technologies to turn crisis into opportunity when faced with the adversity of the Pandemic of 2020.

We unexpectedly confronted the realities of our imagined digital existences during the pandemic. Some of us were better prepared than most, but even we were forced to improvise to confront unexpected circumstances during social isolation. Trying to predict, much less arrest, technological change has always been a fool's game, never more so than today.

Many of us have lived lives spanning multiple paradigms without realizing it. I can still remember the thrill of going to see Star Wars in the theater. I also remember the thrill of being able to interact with Star Wars in a game on my computer in the 1980s. Now I watch the latest installment via Disney+ with almost no transactional friction. I remember spending hours in record shops and bookstores looking for treasures to nurture my soul. Now, I can download movies and books instantly as well as subscribe to an endless buffet of music through streaming services. These transactions have essentially become frictionless. However, the data they generate is also frictionless. Sure, the record store could have kept track of the music or movies that I was purchasing, but it was an extremely inefficient process to share that data and the store had little incentive to do so. Amazon, Facebook, Google, and Apple have easy access to my purchases across multiple platforms and ample reason to share them.

Digital technology has made all kinds of information fluid. Fluid means transparent and that inevitably raises concerns about privacy. It's a lot easier to find stuff out about you if it's online somewhere than locked in a dusty filing cabinet at the county courthouse or your doctor's office. However, there is also a valid argument against outdated notions of privacy. Thinkers, such as Jeff Jarvis, have made strong cases against random laws that restrict the benefits of sharing data. (Jarvis, 2011) For instance, even in the most sensitive areas such as our medical data, we have found cures and treatments because digital affordances allowed the unprecedented sharing of data. Data has been critical in the fight against Covid. However, it is also data sharing that allows the Russians to hyper-target voters in elections that they are seeking to influence, peddling targeted, often contradictory, messages to sow chaos.

These examples show that technology is not moral. It just is. We can no more regulate its morality than any other aspect of it. That's a little like blaming a whorehouse for the activity that goes on within its walls. We can, however, demand that people and institutions obey the cultural norms of the physical world. We do not run naked into the public square. We should not do that in the virtual public square either.

Culture expresses the reality that humans are creatures of habit. It wasn't that long ago that our very survival depended on being effective pattern recognizers. When we are taken out of familiar paradigms, which are nothing more than amalgamations of patterns, we become profoundly discomfited. This is as true as when we had to adapt to shifting from the dangers of the forest to those of the savannah as it is when we have to shift from the dangers of an analog, industrial world to a digital, increasingly re-combinatorial world. For instance, propaganda is not a new thing, but in an industrial world, we knew that publishing information in a newspaper would be spread to all who read it. As a result, that publication would develop a reputation for neutrality or different flavors of bias. Now digital technology allows Facebook to be simultaneously everything to everybody, because you never see what you don't want to see. Others don't see what you see either (unless the algorithm says that they do). This can be profoundly dislocating to those who are used to the legitimacy derived from older paradigms of newspapers and broadcast television.

News is just one symptom of this kind of shift going on. Media is merely the first part of the wave that is engulfing human society. Medicine and all forms of science have been profoundly changed by relatively frictionless data transfer. Physical items are taking on the characteristics of digital ones as 3D printing and CNC (computer numerical control) — driven shop machines are becoming increasingly common. If you can create a digital design through simple design tools, you can create a physical manifestation of it. Cars are likewise becoming modular as the need for form to follow function is eroded by replacing complex gasoline motors with simple electrical ones and mechanical systems with electronic ones.

These factors point to an ever-increasing rate of change. However, there are many, many sectors of business and the economy that still operate at pre-digital speeds. The law is one good example. If you want to have a time warp back to 1975, get involved in the legal system. Lawyers are some of the most technologically resistant people I've ever met. The courts themselves have resisted technologies that would make their proceedings more transparent to the societies that they serve. However, many industries are not far ahead of these institutions. Note, for instance, the continued widespread use of DOS-based cashiering systems in retail or even the use of tractor-feed printers in certain industries.

In the case of regulation, it's clear that most of our political class have little or no understanding of the new realities of information and data flow. Existing bureaucracies are poorly designed to handle the new realities of technology and information. Witness the Federal Communications Commission's struggles with Net Neutrality and other new media questions. Fixing these kinds of issues requires a legislature that understands how digital transformation is profoundly changing the landscape as a host of unimagined questions emerge. For example, copyright and patent law have struggled with intellectual property issues in an environment characterized by the easy sharing of information.

As this sharing moves into the physical realm through exchanges of 3D printer designs or, at the other end of the spectrum, companies' attempts to create artificial monopolies through the use of chipped consumables like water filters and toner cartridges, it's unclear how human systems will respond. These kinds of sharing do not respect national boundaries. If you figure out how to stop copying here (and that is a dubious proposition), it will move to another legal jurisdiction.

Through all of this, we often seem to lose track of our basic human tendencies. Technology may change rapidly, but humans don't. We are still struggling with the same foibles, blinders, and habits we have been struggling with throughout recorded history. Cultural ignorance leads to opportunities for corruption and other age-old human failures. Technology accelerates that. It stresses our institutions in ways for which they are poorly designed. Economic interests will often recognize opportunity in these failures.

Technology is never immoral, but humans can be. Technology is always amoral. It's the uses humans put those technologies to that apply a moral imprimatur onto a technology. This is what we are seeing when we see the "technology" failures of the current era. These are humans leveraging their tools for good or ill. The problem is that we often lack the skills necessary to peer beyond the technology to see the human tendencies underlying them. We have to develop mindsets and skill sets that permit us to perceive those kinds of changes that are positive and those that are negative. We have to recognize that human organizations, from businesses to schools to bureaucracies, will often apply technologies to serve reactionary needs. ❈ STRUCTURE is almost always put into place in order to insulate processes from change. Digital change can be profoundly threatening to those dedicated to preserving the status quo.

Any discussion of the evils of technology must therefore start with a profound evaluation of the human activities that underlie it, for it is there that you are most likely to find the evils, ranging from wasted potential to outright murder. How do we design technology to take advantage of huge positives it holds for human society? How do we rethink teaching and learning in order to prepare ourselves and the coming generations to adapt to the level of technological change that is occurring? How do we rethink how our businesses work in order to surf this wave as well? How do we create cultural change that is evolutionary rather than revolutionary? How do we find our digital selves? These are all central questions that we must address in order to stave off disasters of apocalyptic proportions and to create positive futures for ourselves and our children.

Technology is speeding up all of these processes. That is why we need to pivot around the human aspect of these problems and not the technology surrounding it. Until we solve for the human, the technology will always win. Solving for the human involves an evolution away from rigid industrial mindsets toward fluid digital ones. This means we need to rediscover our childish explorations and creations and not allow industrial processes to turn us into unthinking machines.

Industrial-era organizations of all stripes are dehumanizing in their structures. This was technologically driven, as industrial norms required humans to become cogs in various industrial machines, from the factory to the school. All operated on principles of efficiency that often stifled the very humans that they were meant to provide for in order to scale their benefits. Mass production, mediocrity, and adherence to the norms of the group subsumed craftsmanship, artistry, and individual creativity.

Over the last 20 years, technology has gradually liberated us from the necessity of these kinds of organizations and structures. However, they are still with us and, instead of maximizing efficiencies and progress toward solving the problems of society, they have often been used to dehumanize us further with automated systems that track industrial types of activities more efficiently. It's easy to feel like Chaplin being dragged through the gears of his factory. The gears are just digital now.

We have to recognize that technology is often used to make outdated processes and paradigms more efficient rather than being used to augment us as humans. In an open market, these strategies will be increasingly untenable as digital firms rapidly outstrip those stuck in industrial modes of thought. Innovative organizations harness human creative potential, not human mechanistic potential. The rules have changed. We're just now understanding how our paradigms of achievement, organization, and a life well spent have to change with them. This book provides a roadmap to lead us out of the factory and to reimagine what we are capable of as humans.

We almost never approach the ideal of artisanal work in our daily existences. While I am not suggesting that every computer user must learn to code, I am suggesting that everyone needs to reflect at some level what the connection is between what they are doing and how the technology augments or impedes that task. We would quickly discard a hammer that doesn't drive nails and yet we often find ourselves trapped inside of digital tools that are equally useless.

If circumstance forces you to keep using a poorly designed digital hammer, it is fundamentally dehumanizing. ⌛ **TIME** is the true non-fungible commodity. You can never get it back and pursuing pointless, hopeless work is not what we were placed on this earth to do. It is a simple question of empowerment. If you do not know what the purpose is in the work you are doing, you are essentially a slave. If you can divine the larger purpose behind your activities, then you can be a master of your environment.

The Industrial Age provided few opportunities for the individual to become master of his or her own environment. The Digital Age does. Our paradigms of work and life have shifted even if few realize it yet. We can all become artisans if we understand the impact of our particular art on the world. This is true on a personal level, as we construct technological environments in our homes, create digital personalities, and create personal artifacts in a bewildering array of forms. It is also true as we consider our business, governmental, or educational organization's role within a digital ecosystem. Technology can empower us to become agents rather than simply being tools.

What shocks me is how many choose to remain tools in this dynamic environment, full of rich possibilities. The reasons for this are complex, but they are almost entirely human and cultural. They reflect a series of choices, often based on outdated assumptions, that we make about our personal relationships with technology, and, by extension, our organizational relationships with it. Our educational systems also acculturated us to it from an early age. Education is also an industrial paradigm. Students and teachers are widgets as much as anyone working on a factory floor. The product is people, not cars, but the process is often remarkably similar.

Those that can make the mental transition to thinking about communications, production, and work in new ways will have a decided economic advantage. Those who do not will find themselves increasingly out of sync with an environment characterized by rapid technological change. Even if we are willing to accept being the Little Tramp stuck in the factory, this is no longer tenable. Machines will replace that kind of activity, whether it is physical or mental. Climate change is also going to alter the consumerist realities that underpin more and more production. The industrial world as we know it is going away. It is unsustainable for our planet. What role we have in shaping the Digital Age depends on how effectively we grasp digital thinking.

The consequences of remaining stuck in the wrong technological paradigm will be increasingly grave for our social institutions and our competitive position in the world. We must choose to maximize human potential through our technology. Dehumanized people are a recipe for finding vast swathes of our workforce replaced by the machines. This intent of this book is to provide a path to remapping a human connection with technology that industrial thinking has profoundly warped. Its intent is to re-humanize our relationship with change and the technology that drives it.

Managing nonlinear change will be the human project of the 21st century. It is only through understanding ourselves that we can successfully bridge the gap between *homo industrialis* and *homo digitalis*. It is only by stepping back from the technology itself and asking fundamental questions about why we do certain things, why we organize to do certain things in certain ways, and why we accept the nature of our technological environments, that we can address the challenges of an ongoing digital transformation.

Transitioning to the Industrial Age cost millions of lives. It was a slow process of almost two centuries (and some countries never fully made the journey). It has now run its course. However, the transition to the digital world will take decades at most and may even be substantially completed within a single decade. Humans, their organizations, and their cultures have never shifted paradigms that fast. Thomas Kuhn argues in *The Structure of Scientific Revolutions*, published in the same year as Engelbart made his proposal for augmenting human intellect, that humans adapt to new paradigms slowly, over years, if not decades. (Kuhn 1962, pp. 160-173) We may not have the luxury of that much time. We can no longer afford to be the Little Tramp.

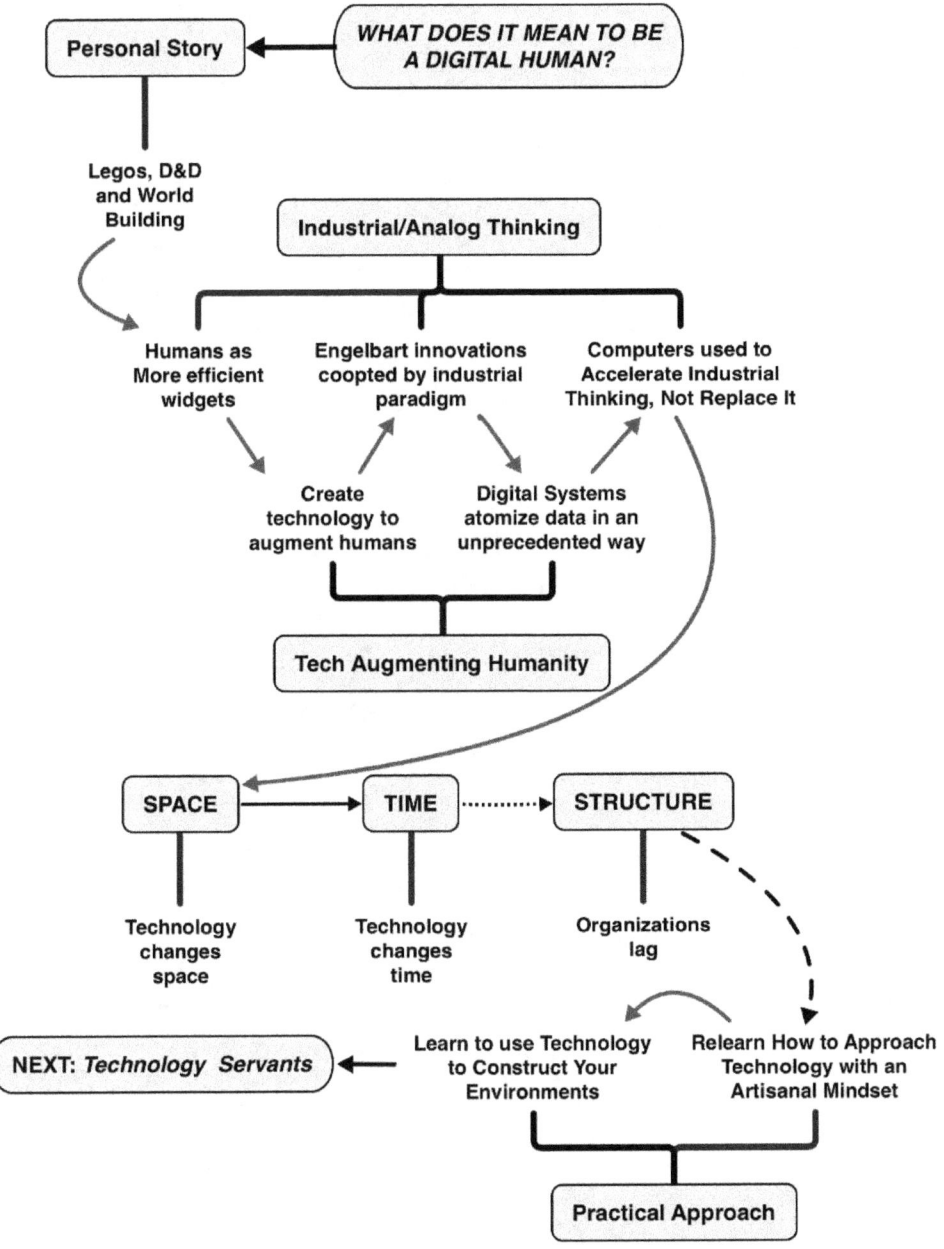

1.0 Designing Human-Centered Technology Systems -

A brilliant solution to the wrong problem can be worse than no solution at all: solve the correct problem.

Don Norman

✪ SPACE

We rarely think of ✪ SPACE as a technology, but we constantly create spaces with our various technologies. How we approach our technology environments is critical to our ability to "think digitally." Creating a productive relationship with it is a necessary first step to finding our digital humanity.

1.1 Technology Servants

We refer to a way of life in an integrated domain where hunches, cut-and-try, intangibles, and the human 'feel for a situation' usefully coexist with powerful concepts, streamlined technology and notation, sophisticated methods, and high-powered electronic aids.

Douglas Engelbart, 1962

I was driving down a country road in New Mexico with my best friend. We were hunting photographs. It had already been a productive day, with amazing black and gray clouds over White Sands National Monument and dramatic banks of cumulus kissing the Sacramento Mountains. Then the road took us west toward our overnight stop in Socorro. Driving along US 380, the sun was setting into a black curtain of clouds along the western horizon, making for a dramatic scene. We stopped by the side of the road to capture that scene, and I would get a dramatic shot of the road disappearing into darkness.

It was, however, when I turned to look back south that the vista that opened before me made my creative vision leap. A dirt road disappeared into the desert, but above it there was a swirling miasma of ever-changing clouds. In an instant, I saw the shot in front of me and quickly placed myself and my camera in position to get it. There was little time to adjust all the controls on my camera, but I knew just what lens and exposure setting to use. I saw past the scene in front of me, through my tools, and to how I could reproduce the image that I saw in my mind. At that moment, I achieved mastery over the technology. All that was left was my struggle with the art.

People often ask me how I do the photography that I do. It is a mystery to them. I admit there is a certain amount of innate talent involved. I am predisposed to recognizing patterns of all types. Ultimately, however, my interactions with photography have always involved how I approach technology. I am not a "gear head." I am not constantly fascinated with the latest and greatest. The camera is merely a means to an end. The end is translating my mind's eye into the final image. I usually have a pretty clear visualization of a picture before I ever trip the shutter. Getting from idea to image is a journey through technology. Every technology in between, from the darkroom to Photoshop to print services, is a potential barrier and my foremost goal is to chart the most efficient pathway to achieving the final image. I get frustrated when something is not working for me and want to understand why. In this way, technology is all about simplification and demystification for me. Complexity is an anathema to my goal.

I learned as a teenager that understanding technology could allow me to reduce complexity in my interactivity with communication of all types. It was another quarter century before I could combine this with my photography, but I could see the direction that things were taking even then.

In 1983, while I was a teenage member of the Houston Area Apple Users Group (HAAUG), Steve Wozniak came to speak to one of our meetings. His talk was all about his early days as a hacker at Berkeley. During his time there, M*A*S*H had become a big hit among the students and they used to gather every Monday night around the only TV in the dorm to watch it. Woz had developed a TV jammer that would send out signals that would scramble the set from the back of the room. He used to revel in making the students have to get up and mess with the antenna, thinking that his jammer was actually just a poor signal. He would wait for them to get into the most awkward position possible and then turn off the jammer, thereby fixing the "problem." One night, he effectively trained one student to stand on top of a stool with the antenna held up as high as he could reach, while a second student could not remove his hand from the screen. The group was forced to watch the rest of the episode while performing these gymnastics.

From the perspective of the intervening decades, I understood Woz was playing with his fellow residents' relationship with technology. He had, through almost Pavlovian means, trained them to do the most illogical and outrageous tricks. He was a master of the tool. They were the servants. And he quickly recognized how to take advantage of that reality.

This does not have to play out this way. We can all learn to be like Woz (and he would be the first to encourage you to do so). One reason I am an effective photographer is because I am the master of my tools. Like Woz, I can hack the reality of the world using the technology available to me. This is how everyone needs to approach the technological world of the Digital Age. As one of Woz's contemporaries, David Rodman said in the early 80s:

> A really good program designer makes an artist out of the person who uses the computer, by creating a world that puts in them in the position of "Here's the keyboard, and here's the screen. Now once you learn a few rudimentary communication skills, you can be a superstar." (Rheingold 2000, p. 173)

One of the chief goals of this volume is helping humans harmonize their tumultuous relationship with technology into mechanisms for the creative hacking of our personal, organizational, and societal challenges. We must figure out how to find the same harmony with our machines that I found with my photographic tools on that day. Humans face an ever more complex world driven largely by technological change. We cannot alter the rate of this change short of an apocalyptic disaster. What we can change is our relationship with technology. We can either embrace our tools or find ourselves overwhelmed by them.

Consider for a moment all the daily decisions you make, both at home and at work or school, and how you could streamline your approach to making those decisions. Over sixty years ago, J. C. R. Licklider said, "my choices of what to attempt and not attempt were determined to an embarrassingly great extent by consideration of clerical feasibility, not intellectual capability." (Licklider 1960, 2003, p. 76) This reality still shapes to a great extent how we approach problem-solving, from a personal to societal level. Instead of using technology to simplify our tasks, we have used computers to increase the complexity and speed of "clerical" activities that we must confront. Instead of giving us more ✪ SPACE for creative intellectual pursuits, we have created more dehumanizing barriers. I see time and time again the behavior of Woz's dorm residents. People are constantly doing gymnastics in the service of badly designed, "magical" technology solutions. In this kind of world, we are the servants, not the technology.

This does not have to be. We all can hack our way out of this technological thicket if we recognize what is going on is at its root a human failing, not the failings of some sort of technological "god." As my photography example illustrates, I transcended the technology of my camera when I was creating art. We don't think about the pencil (unless it breaks) when we are writing. Writing absorbs us, not pencil operation. Mihaly Csikszentmihalyi calls this state "flow" in which the activity you are undertaking absorbs your entire consciousness toward an intrinsically rewarding purpose. (Csikszentmihalyi 2008)

We often perceive technology as impeding our ability to achieve this state, but this is often a product of poor design, lack of user empowerment, and a confusion of ends over means. Our tools get in our way as we try to create our way out of challenges of both the artistic and mundane variety. For a long time, this was because those tools were too complicated. Alan Kay said, "for most people, the piano has been the biggest thing that turns millions away from music for the rest of their lives." (Kay 1986 at 44.22) The piano is an immensely powerful tool. It vastly simplified what came before it. However, as a technology for making music, it is still served by a relatively small elite who can invest the ⧖ **TIME** to master it. It locks out many others. It is only through thousands of hours of effort that someone can control the piano to a level where the music can overcome the technical difficulties of operating the machine designed to create it. Most of us don't have the time or devotion to devote thousands of hours to mastering expressive tools, but that doesn't mean we don't have musical ideas to express.

Digital Age technology makes new kinds of tools accessible to many more people by lowering the barriers between ideas and their execution, but there is room for so much more here. The doors are open. The opportunities are there. We just have to grasp them by demanding better tools.

For much of human development, machines did not figure into our mental existence in the same way as they have done over the last two centuries. Tools were an adjunct of our natural environment and had been since the dawn of recorded history. They extended our muscle strength and allowed us to turn soil, construct homes, hunt, and kill one another in a more effective manner than our bare hands. Tools, however, did not overwhelm our natural existences. Most humans farmed and tools were nothing more than a way to manage nature.

During the 19th and 20th centuries, the industrial environment gradually replaced the natural one in many parts of the world and with it came a prevalent sense of learned helplessness in the face of The Machine. This trend sped up significantly following World War II. Instead of using machines to tame the natural environment, we have used them to replace our connection with it. Instead of machines connecting us to the world, we became subsumed into the machine itself, as Charlie Chaplin so aptly parodied in *Modern Times*. Henry Ford's assembly line enshrined the worker as merely a cog in the machine. We have since extended that to realms far beyond the factory into the information industry. Unsurprisingly, the word "robot" also originates during this period. It first surfaced in 1920 in Czech in R. Capek's play *R. U. R* ("Rossum's Universal Robots") and literally means "forced labor." In the world of Ford and Chaplin, humans *are* the robots. We are chained to the factory machine. It should come as no surprise that humans grew to think of themselves as prisoners in this system.

The new mental machines that emerged after World War II had the potential to liberate us from the industrial shackles that our tools had become. As Douglas Engelbart wrote as early as 1962, "We refer to a way of life in an integrated domain where hunches, cut-and-try, intangibles, and the human 'feel for a situation' usefully coexist with powerful concepts, streamlined technology and notation, sophisticated methods, and high-powered electronic aids." (Engelbart 1962, 2003, p. 95) Engelbart's conception of our relationship with technology contrasts starkly with that of the factory floor (or even the 1960s office environment) where machines essentially enforced conformity with the systems that were necessary to sustain them. This is counter to what Engelbart envisioned technology being able to do. In his vision, technology didn't enslave us, it empowered us. He and many of his contemporaries such as Ted Nelson, J. C. R. Licklider, and Seymour Papert, to name just a few, showed us the door. Most of us have not walked through it.

I have been directly involved with computing technology for almost 40 years now. For the last decade, many of my projects focused on designing and facilitating the transition of education from industrialized shackles into the digital era. As someone who is always looking for tools that will allow me to do new things, the pace of change has often been frustratingly slow or characterized by long detours down unproductive pathways. For the last 25 years, we have been predicting that the internet will provide that liberating force and periodically have had tantalizing glimpses of what that might look like.

The impact of the COVID-19 Pandemic and its sudden separation of all of us into discrete pods through social distancing, remote work, and remote learning forced most of us to grasp at unfamiliar tools. We found many tools to be wanting. We found in others unexpected utility. All of them forced us to take a hard look at what we were doing. However, what became clear was that the vision of the internet pioneers has yet to be fulfilled. We chained students to lengthy Zoom lectures. We forced students and employees to install spy software on millions of computers to make sure they weren't cheating and that they were working diligently. Instead of exploring the possibilities, far too many found themselves in a technological panopticon. This only reinforced the perception of the relationship with our tools through the lens that Charlie Chaplin was mocking in *Modern Times*. Instead of liberating us, the computer replaced the drill and wrench as our master. Only now technology made it a much more efficient overseer.

My good friend and colleague, Dr. Bryan Alexander, made a point in a talk over a decade ago about how humans adopt technology. He said it always happens in two phases. In the first phase, the technology is seen as a more efficient but analogous version of an existing technology. Hence, the automobile was a "horseless carriage" first. In the second phase, the technology reshapes what *we* are. There is no doubting the impact that the automobile had on the 20th century, from highways to suburbs to warfare. Likewise, in the 1980s, most businesses jumped into the first paradigm of PC computing and replaced their punch cards with databases; their typewriters with word processors; and their calculators with spreadsheets. However, with very few exceptions, they never transitioned to the second phase.

There is a direct line between the drafting machine and AutoCAD, interoffice memos and email, the typewriter and the word processor. Direct lines are conceptually easy. It's no accident that the first printed book in Europe was a Bible, something which had been carefully reproduced by hand in monasteries for centuries prior to Johannes Gutenberg's press in 1453. Arguably, it took decades before the capacity of making exact duplicates of information on a large scale impacted how we connected and shared knowledge significantly. The technology was there. Humans were the lagging indicator.

We still have a poor understanding of how the Internet will impact ideas and tools and have to realize that we are still in the very early phase of human adaptation. We are still duplicating the modalities and processes of the Charlie Chaplin/Henry Ford world just using digital technologies to do so more efficiently. What we need to realize is that we won't be printing our equivalent of bibles (such as the interoffice memo) forever (or for very long). We are just beginning to explore how the affordances of mashing up ideas in new and previously impossible ways will change human existence.

Rethinking everything is as much a structural problem as a psychological one. Structurally, our organizations have grown up servicing technology, either directly or indirectly. As a result, we have attracted many people who see the technology as an end unto itself rather than as a means to an end. Their function, a necessary one, is to get the technology working, and it ends there. This is more easily accessible because it has cognates in older paradigms. It's just a more complicated version of a telephone technician or even a plumber. However, this vision sees technology as an endpoint rather than a starting point for new explorations because it is a first paradigm viewpoint: the internet is the same as broadcast, the computer is a fancy typewriter, and social media is just the postal service. As Douglas Engelbart and two colleagues pointed out in a 1973 talk:

> Workshop improvement involves systematic change not only in the tools that help handle and transform the materials, but in the customs, conventions, skills, procedures, working methods, organizational roles, training, etc. by which the workers and their organizations harness their tools, their skills, and their knowledge…. Development of more effective knowledge workshop technology

> will require talents and experience from many backgrounds: computer hardware and software, psychology, management science, information science, and operations research, to name a few. (Engelbart, Watson and Norton 1973, p. 9)

These kinds of paradigm shifts aren't limited to what we now call Information Technology. They impact more traditional fields like architecture as well. In architecture the technology is the built environment. Technology has stretched what architecture means both horizontally and vertically. Horizontally, increasingly complex ecosystems of buildings with their natural environment, with new technologies, and with other buildings have made each building project far more complex. Vertically, the concept of the building as a living organism has been brought to the fore, as technology makes it much easier to reshape the built environment dynamically. The attendant recognition of the inter-relationship between the built environment and the communities that inhabit it is much harder for those used to product-focused, closed processes to wrap their heads around.

We can construct these same kinds of scenarios for most industries once you drill down to the ends of accepted functions and get past the industrial means used to get there. The Industrial Age was about taking things apart. It was about specializing. The Digital Age is about putting things together and creating opportunities in the intersections and edges of what were originally specializations. The Industrial Age manifests itself in many ways. Education is focused on graduation and we treat students as widgets being produced by the system. Governments ignore the interconnectedness of systems as they go about regulating and deregulating parts of the economy. Industrial business creates ever more specialized and complex products. There are powerful forces throughout business, education, government, and even technology, maintaining the status quo even as the fundament shifts beneath them.

What does this mean to the average user of technology? It means that systems construct structural barriers to using technology effectively. These significantly limit how much any technology can augment an individual's productive and creative output. We often mislabel structural barriers as "technological" problems.

Simultaneously, the media often portrays technology as a threat to be managed rather than recognizing the human and societal problems that a particular technology often unmasks. We don't have more crime (it's actually down since the 1970s). We can just see it better because of technology. Video games don't cause mass shootings. They're just a lot more visible than the imaginary fights that we used to have by shooting at each other with toy guns. This culture of fear has created whole industries of middlemen that are ostensibly there to protect us from the evil magic of technology but, ultimately, we discover that what they are protecting us from is ourselves.

The machines themselves provide us with perceptual challenges as we struggle to adapt to their existence. For the first 30 years of the PC revolution, as computing technology saturated our lives, they struggled with reliability, much like early cars did. This resulted in the need to develop a core of middlemen whose mission was to protect us from system failures, viruses, network incursions, and so on. This is a limited job. Its goal is one of boundaries, setting them, protecting them, and enforcing them. I understand the need for this. I've done this job. It's a lot easier to do when you can control the environment. However, this focus perpetuates industrial-era notions of servitude to the machine and blinds you to its opportunities.

I find myself frequently at cross-purposes with my brethren in the pure Information Technology world, because I look at technology as a human enabler first and foremost and, more often than not, this challenges some rigidly structured environment. I get into arguments with them about human-centered design and the fallacy of ignoring human behavior when designing technology systems. In their conceptual framework, humans are something to be designed around in service of the machine. Like in Henry Ford's assembly line, humans must adapt to the needs of the machine, not the other way around. This perception creates many problems for humans searching for ways out of the maze of technology and for the opportunities it should open. Badly designed technology is a barrier, not a tool. Technology should create "opportunities," not "solutions." If we portray technology as a "solution," that implies an end point to development, creates barriers, and may be perceived as threatening. The benchmark for good technology is that it should always strive to simplify tasks for the broadest possible set of users, not make those tasks opaque and unmanageable.

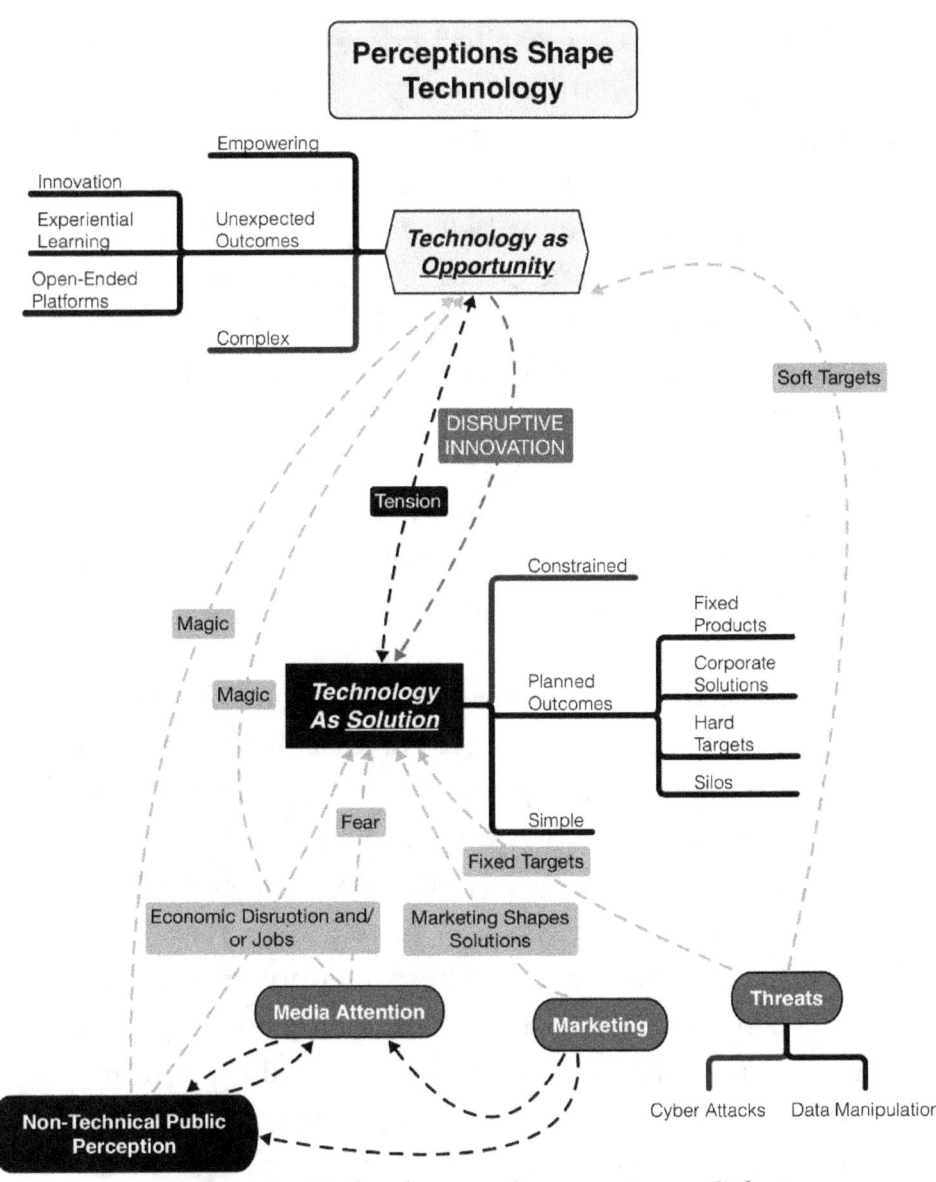

Figure 2 - Technology as Opportunity vs. Solution

There is no excuse for technology systems in the digital age to be hard to manipulate. Industrial-era analog technology was hard because it often involved physical manifestations of muscle. Large inflexible machines, from conveyor belts to spot welders, were difficult to adapt to the circumstances of the human, so the human had to adapt to them. Digital technology is more accessible because you can *customize* the technology around the needs of the human, enabling opportunities for the augmentation that Engelbart describes. Technologists always view technology systemically and are constantly hunting for opportunities to remix technologies to further human creativity.

The Digital Age requires a fundamental rethink of how we approach technology. Historically, we design our tools to fix specific problems. If you need to saw wood, you created a saw. If you need to hurl projectiles at your enemies harder than you can throw, you create bows or firearms. We need to go over there. We will create the wheel. These technologies create other vistas, but we often fail to perceive the much larger impact of those opportunities. Sailing is much easier than rowing and we can get large amounts of goods up the coast with a ship blown by the wind. However, we can also explore whole new continents and perceptually change the shape of the entire world with these sails as well. They are paradigm makers, even as they are paradigm breakers.

Many of our problems today stem from a disconnect between an analog mindset that perceives tools as closed solution sets and the reality that true digital technology should not behave in this way. The media narrative about technology is almost always about solutions and the reactions that they cause. Robots are perceived as a solution and a threat. We paint automation as a solution and a threat. And, in an implicit conflict of interest, the media likes to portray Facebook in threatening terms. Solutions attract attacks, and this also helps fuel a media narrative of technological chaos. Hackers and cyber security breaches are seen as the inevitable consequence.

The modern media version of hacking (going back to the 80s) started because kids wanted to attract attention by penetrating the Pentagon, a major corporation, or other "hard" targets. As the value of these exploits has increased, money or power have taken over from the show-offs and are far more dangerous. Our fascination with solutions has led, in part, to the seriousness of the situations we are now confronting. It also leads to a lot of fear, which the media gladly takes up because fear is an easy way to get the attention of viewers.

Analog solutions produce castles. The castle was a natural technological solution to the necessities of the agricultural human. At its apex during the medieval period, the castle was also the ultimate expression of hierarchy for its age. There are three problems with castles (that are shared with all technological solutions). First, they fix you in place and provide a focal point for any attacks. Second, castles lead to technological and social stasis. The lord of the manor has little incentive to change the society that has put him in his position of power. Therefore, there is a powerful incentive to innovate just enough to manage perceived threats and quash down on technological innovation that threatens that position. Innovation threatens hierarchy. This leads inevitably to the third consequence: eventually someone will make the technological leap to inventing artillery and the entire edifice will become obsolete overnight. True hackers don't just attack the system, *they break it*. This same fate can easily befall any technological "solution." It is only with true digital solutions that we may have finally found our way out of this endless arms race.

Digital technology doesn't behave in this way. It is cheap and has no fixed outcomes. As Claude Shannon theorized in 1948, we can digitize any piece of information. Once we do that, it becomes endlessly re-combinatorial. (Shannon, 1948) With the advent of technologies such as 3D printing, we can extend this information into the physical realm. The consequences of failure drop to nothing, and experimentation and iteration become possible on a previously unimaginable scale.

Digital thinking almost inevitably leads to networks of thought. Ideas are combined and recombined into completely new structures and open-ended opportunities. Perhaps the most visible example of this today is the work of Elon Musk. Is Tesla about disrupting the car business or the energy business? Perhaps both, perhaps neither. Is Space X about getting us to Mars or reinventing the internet?

Again, perhaps both, perhaps neither. Musk may or may not know the answers to these questions either, but what he understands is that he can afford to fail at a scale that would have bankrupted Henry Ford and therefore shoot for the unexpected outcome. The Space X and Tesla projects form a networked web of ideas that will lead to unexpected opportunities and, between them, are likely to create world-changing technological opportunities. They could also intersect in some not-yet-imaged way. Musk is hacking the system, not the technology.

Digital thinking and its view of technology as opportunity harnesses the power of distributed communities. If you want to hide your treasure, you don't put it into a castle; you put it in an empty grave in a vast cemetery like in *The Good, the Bad, and the Ugly*. Blockchain is the digital version of this approach to security. The strength of your community lies in its diversity, not its homogeneity, as it did under the lords of the castle. The strength of technology likewise lies in its diversity. More diverse solutions are more antifragile, as threats to one kind of technology are less likely to be threats to technology with different assumptions built into it. For instance, during the Great Texas Freeze of 2021, over-reliance on one technology source (gas powered electricity generation) was a large contributing factor to the disaster that caused widespread power failures. In elections, having a paper trail to back up the electronic record is considered the gold standard of election security. Those are just two examples where technological diversity increases the resilience of human systems.

Decentralized knowledge is not a new concept. Science has always been based on this model and often operated in tension with the industrial model it helped to create. The distributed "republic of letters" drove The Enlightenment and is a miniature version of the listservs, blogs and Twitter of today. It is also a close analogy to communications technologies ranging from Slack to Internet2. Science has built us castles unimaginable to a medieval lord, but in parallel it has created the seeds for the ultimate destruction of those castles in the universal hive mind of distributed communication and imagination. The latter will be far more powerful than any castle in the end unless the forces of hierarchy can figure out how to strangle it (and I doubt that is possible). It is possible to use these tools to mislead and distract the public, as we will discuss in Chapter 5. However, as I will argue, we can overcome misinformation by requiring high levels of transparency, a natural bias of digital information.

The exciting thing about this version of technology is that it is infinitely customizable, according to the needs of the moment. If you need a digital factory, you just build one in your living room or, better yet, farm it out to a multitude of cloud services. Decentralized production extends way beyond the digital realm itself. Making has significantly lowered the barriers to entry to constructing whatever you need using digital tools coupled to physical output devices. If you need a door sensor, you can build a door sensor. If you need a data platform tracking the movement of people in your building, you can build this. If you need a video wall, you can build a video wall. ✿ **SPACE** and ⧖ **TIME** do not matter in the same way. Digital technology allows us to iterate the world at a pace that was not possible before.

However, this aspect of digital technology threatens those trying to protect their technological castles against what they perceive as disruptive threats. This attitude is completely understandable. If technology is a solution, it demands an analog mindset to protect it. Threats demand action. This is true. However, the digital world is profoundly non-linear and there are always going to be new and clever ways that people create for exposing and attacking the vulnerabilities of a wide range of "solutions." Intruders can always circumvent a digital system based on walls, because this approach ignores the power of automated repetitive attacks. Dedicated intruders will always find the weakest link, which is usually the human who leaves down the drawbridge.

One outcome of this is the demonization of hackers and their close association with technology. This misses the point completely. Steve Wozniak, who was constantly hacking the limits of what he could do out of a sense of fun, is a good example of a genuine hacker. That fun led him, in partnership with Steve Jobs, to develop the Apple computer, which was also my first computer. True hackers are not the guys who go after credit cards. Those kinds of "hackers" are just thieves, often using very low-level technology skills to crack electronic locks. Hackers like Woz challenge human systems, not technology systems. The personal computer challenged the orthodoxy that computing was to be reserved for the few and inaccessible to the many. Together, Woz and Jobs figured out how to hack this system and launch the PC revolution. All hackers implicitly understand the human aspects behind the systems and use social engineering to achieve their goals. Thieves exploit human frailties. Hackers exploit systemic frailties. We need more of the latter and must also realize the difference between the two, so often conflated in media narratives.

This is a key distinction because hackers are essentially amoral. Sometimes the systems they attack are immoral and may need to be challenged. Like technology itself, purpose defines hacking. If the goal is to manipulate and disrupt the political and media culture of other countries, as it was with the Russian Internet Research Agency in 2016, then that is immoral activity. And note that what was ultimately being hacked was our social systems, not our technological ones. While we need to think about how we use the technology of elections and understand their true vulnerabilities, the primary target here was our poor ability to think critically about information given to us by unreliable sources.

Hacking can strengthen both our technological and social systems because it helps us understand our vulnerabilities. The reason that the Russian activities were so hard to guard against was that they involved education, not technical patches. There are no technical patches for fake news, innuendo, and rumor. Those have been around for millennia. Hacking the system has often been a force for good. For instance, protest is a form of hacking. Hacking's goal is to force systemic change by challenging specific injustices and spotlighting the weaknesses of the system created by hypocrisy. The Civil Rights Movement in the 1950s and 60s was an exercise in hacking society. Before that, it was a group of hackers that dumped a bunch of British tea into Boston Harbor. Hacking is central to the American democratic tradition. Its capabilities, like many other social institutions, are radically empowered through the use of technology. Hacking is the breaking of chains. If you are a slave, this is a good thing.

Where things go terribly wrong is when we associate hacking with technology rather than the human drives behind it. By misdirecting the blame for our shortcomings on the technology, we create fear and barriers to access them. The predictable reaction is to double down on what's perceived to have worked in the past and try to restrict human activity. This is not a good long-term strategy at either the societal or local level.

Technology can show us the vistas of the world or we can turn it into Jeremy Bentham's Panopticon. The tools we create can be made available to all to create the unexpected, or we can lock them down and render them inaccessible to limit access to that ability. Passwords are good example of trying to control an uncontrollable environment with outdated thinking. If you think of technology systems as castles, you quickly realize that there are many doors to the keep. You must give the users of the system the keys to the castle, but you want to control that as much as possible. If you turn humans into keys, the results are predictable.

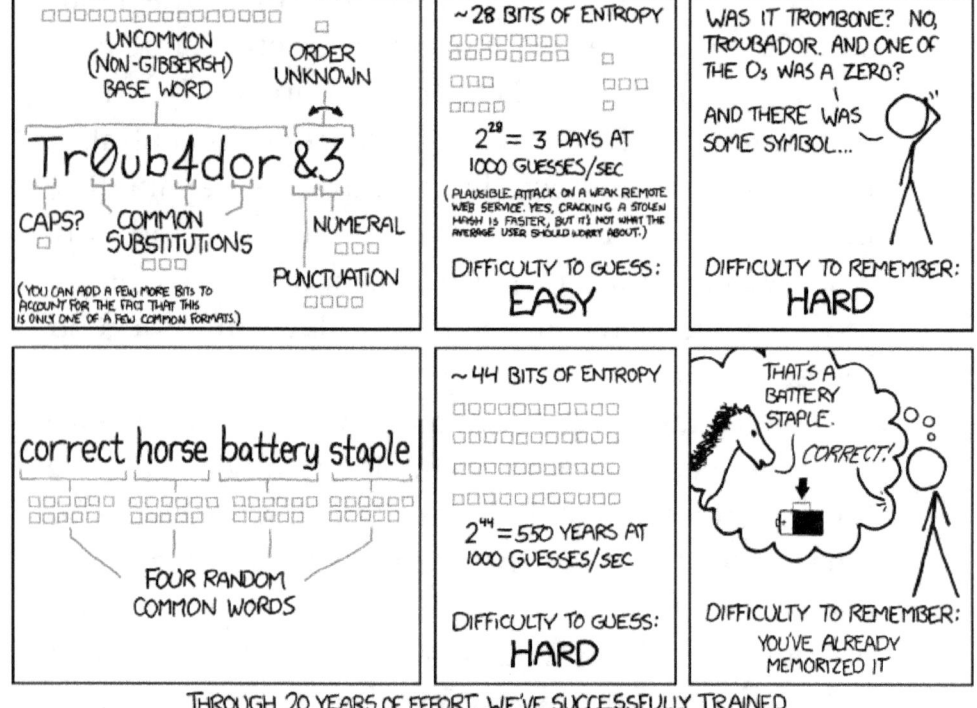

XKCD Cartoon on Password Strength

There are extensive pages devoted to the mathematical proofs of Randall Munroe's argument. Technicians, being focused on technology, fail to recognize the impact this has on their users and their use of technology. Forcing users to change their passwords to some gibberish every 120 days without taking the human element into account creates a lot of barriers to productive use of their tools. It forces users to be slaves to the technology, and even worse, makes it a capricious master. These supposedly secure "systems" have created a culture where many users, unable to remember a constantly changing stream of gibberish characters, tape them to their computer somewhere. Now, you've got a security hole that enables thieves while keeping us from accessing our tools. It generates a cost of access with little gain in security because it is an anachronistic approach to technology that is amplified by that very technology. It should come as no surprise when humans cannot keep up and are simply dragged along by their chains to the system.

This scenario is a product of being fearful of the technology rather than recognizing its potential. It is a natural consequence of misdirected attention and a lack of understanding of the purposes of the technology systems they intend to protect. The solution is often worse than the threat. This adds fear to the user's experience. What happens if you can't remember this ridiculous password? You won't be able to get your work done. These days, most people's jobs depend on computing technology. However, their jobs are not the technology itself (much less understanding how it works), so technological disruptions, and I see these all day long, are threatening to their livelihood. Humans naturally become reluctant to explore opportunities because they see only barriers and threats.

What happens in these circumstances is that the user starts to view technology in much the same way as the IT professional: as something that has to be controlled or else it will be a threat. Fear is the primary motivator all around. Hacking the system is to be avoided at all costs. Good hacking is frozen out along with bad hacking because of the focus on technological solutions over human ones. This stifles creativity and innovation.

What's really sad about this situation is that it was hacking that got us to where we are now. We would not have PCs without hackers such as Jobs and Wozniak. We would not have the Internet without a hacker named Tim Berners-Lee. All computer visionaries are hackers. That is because undermining the "computer priesthood," as Ted Nelson put it, is the only way to push forward technology and not be its slave. (Nelson, 1974, 2003, p. 304) We need to create more hackers, not less of them. It is by bending technology that we unlock its true potential and unlock ours. It also has the beneficial side effect of creating more realistic threat assessments and constructive ways of dealing with them. Russian hacking should not lead to censorship.

Pablo Picasso said, "Art is not truth. Art is a lie that enables us to realize the truth." Art is hacking. Hacking is art. Technologists are essentially hackers of technology. They see it as a means to realize their vision, whatever that may be. Creatives are, by their very nature, hackers. Composers have been hacking musical norms from Mozart to Beethoven to Zappa.

Visual artists are always pushing the boundaries of what is possible in their various media. My battle with photography has always been to make "real" what I see in my head using the technology available to me. I see that technology as both a barrier and an opportunity. Digital photography has given me the ability to hack this process in ways that I would never have dreamt of when I started this struggle almost forty years ago. The really exciting thing about the digital era is seeing the plethora of amazing things that are emerging from our ability to play with technology.

You don't have to be driven in this direction to appreciate what technology can do for creative or productive expression. Since the advent of blogging, Facebook, YouTube, etc., we have become prolific creators. Studies show we collectively write more than ever before. It's not Tolstoy, but it's clear people have something to say. Making is the latest product of this trend. The creative output of society has exploded because of the digital revolution. Where digital transformation has gone wrong is in areas that are constrained by "solutions." Facebook is more of a "solution" than a creative outlet these days, as it has sought to automate and restrict flows of information throughout its vast network. The users don't control Facebook. Facebook controls and manipulates its users. This relationship is the root of the problematic relationship between Facebook and its users.

The pernicious outcomes of these relationships lead to fear, which the media gleefully covers as Facebook is seen as a competitor for consumers' time.

Facebook is only part of this landscape, however. The media is littered with stories of technological fear: online pedophiles, credit card hacking, Pokémon Go muggings, etc. Media will always gravitate toward the sensational and the sensational is overwhelmingly negative. Negative stories about Facebook, Google, and the other tech giants drown out news about the opportunities that Digital Age tools are opening for us (Allen and Castro 2017). The focus of negative coverage has the effect of closing doors to the genuine opportunities the technology offers us. If we want to harness technology to confront the actual problems of democracy, economic disruption, and climate change, this is no longer acceptable.

We need to get rid of this concept of "technology" writ large because it intimidates many potential creators. The trick is to demystify and exorcise our technology, not to create mystery and fear. As Arthur C. Clarke wrote, "any sufficiently advanced technology is indistinguishable from magic." I think what a gifted carpenter can do with a saw, lathe, and sandpaper is akin to magic, but I have at least an implicit understanding of the technologies he is using to do it. I also see computers in the same way. I don't understand all aspects of programming or chip engineering, but I know enough about the basic concepts to where I can extrapolate. However, I am unusual. Most people view their computers (and increasingly their phones, fitness bands, etc.) as magical talismans. If treated properly, they can do great magic. If angered, they can punish you. This can be a crippling problem. Hackers don't believe in magic. They make it.

Magic inevitably creates priesthoods. This is precisely what Ted Nelson was warning us about in 1974. Magic also enslaves. What we don't understand, we are subservient to. Those who have a more imperfect understanding of technological context become sacrifices to the gods. Doors are closed. Barriers are raised. The environment is controlled. We build organizations with this philosophy in mind because so few in the senior ranks understand technological context and, instead of being inspired by technological possibilities, are afraid of the black magic that they perceive in new technologies that they feel they are being forced to use. Fearing hacks, they simply follow the lead of the professionals in the IT world.

Technology should augment human intellect, in the words of Douglas Engelbart. The beauty of the digital era is that we can all hack our realities. Rigid industrial structures needn't confine us. We can hack boundaries and change the conversations that rule our lives for the better. Digital technology can make this happen if we think about it as a tool rather than as a source of black magic. Everyone needs to become a technologist. We have implicit, if often imperfect, understandings of many systems in our world. If we simply step back and ask hard questions about the tools, we are being asked to use and not simply accept those that serve the few over the many, that will make technology more accessible to everyone, reduce threats, and create opportunities for creativity and growth.

We need to approach these vistas without fear of the technology itself. Our perceptions of technology create mental ✿ **SPACES**. Our construction of technology creates physical ✿ **SPACES**. Both interrelate and create, or impede, human augmentation. Like we have done throughout human history, we need to liberate and regulate where necessary the predictable behavior of humans, not their machines. We should use technology to accelerate creativity and then get out of the way to see what humans can do with it. As Steve Jobs described it, technology should be "a bicycle for the mind." We don't know when the next Jimi Hendrix or Steve Jobs will come along and create magic. Creating magic is always better than worshipping it.

We are all magicians in our own ways. Woz created magic by hacking together the Apple computer. I create magic for myself by taking photographs or exploring new ideas through concept mapping. This should be the fundamental purpose of technology. It should serve humanity, not enslave it. This is our choice, not the choice of the machine. With human-centered technology, there is nothing to stop us from creating all sorts of magic. We just have to decide to do so.

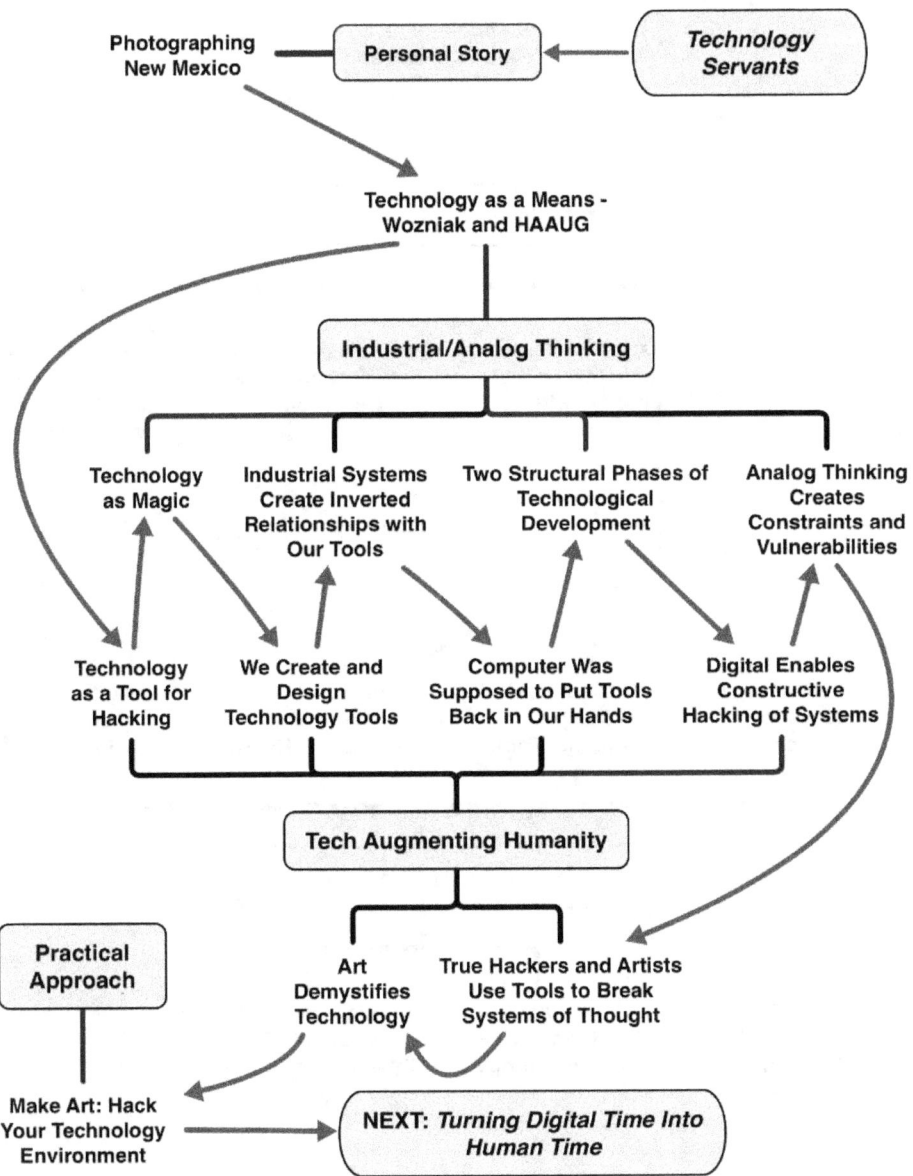

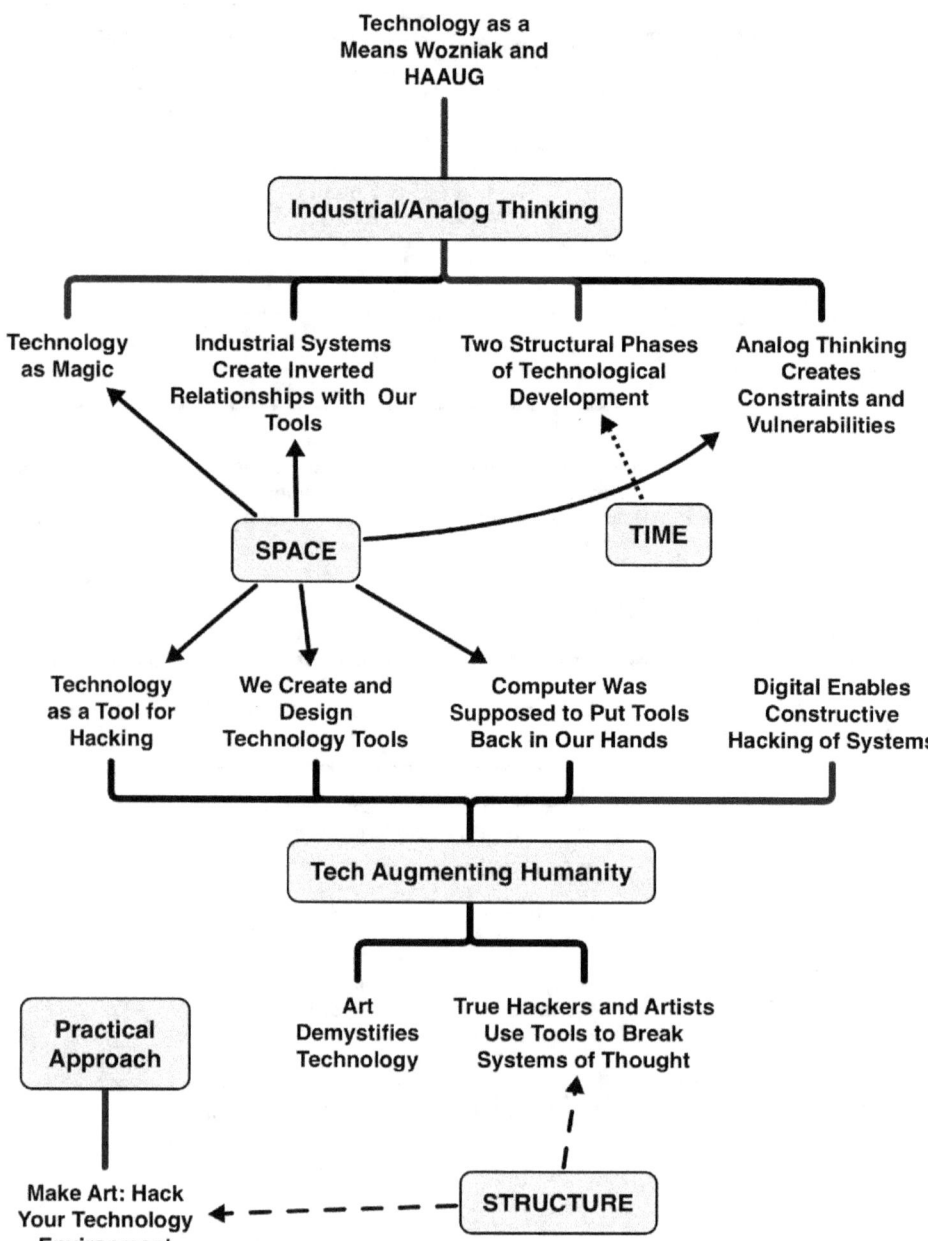

⧖ TIME

The digital environment requires the mindful use of ⧖ TIME. It requires humans to rethink much of what we've been doing for the last two centuries and therefore requires us to stop and reflect on what we are doing. Simultaneously, digital time fragments the notion of linear ⧖ TIME as we perceive it. Understanding how to navigate digital timescales is critical to realizing the opportunities offered by the Digital Age.

1.2 Turning Digital Time into Human Time

Thus we may have knowledge of the past but cannot control it; we may control the future but have no knowledge of it.
   Claude Shannon, 1959

My decisions on what to attempt and what not to attempt were determined to an embarrassingly great extent by considerations of clerical feasibility, not intellectual capability.
   JCR Licklider, 1957

I love deadlines. I like the whooshing sound they make as they fly by.
   Douglas Adams

It is the first day of the semester as I stand in front of my class. As is my normal first day procedure, I ask my students why they were there. The question usually stops them in their tracks because no one has asked them this question before (at least no one expecting a proper answer). The answers come in.

"This is a required class." "I need it to graduate." I tell them that those are obvious answers that focus on process, not intention. Dishearteningly, almost no one ever says, "I want to learn something." I continue to press, pointing out that they do not even have to be in college if they don't want to take "required" classes. The conversation flips toward the utility of college in finding a job, but then they are once again flummoxed when I ask them why that matters.

Why is it that my students have never consciously reflected on their process of learning? It's because our system never really gives them the chance to. Education has become industrialized in its thinking, just like most other aspects of human existence. Society feeds you into the machine at age 5 and spits you out when you can't take it anymore. Some endure to graduate school or medical school, but most don't. The process is not unlike Charlie Chaplin's Little Tramp being fed through the gears of the factory machine. Only in this case, the machine doesn't just imprison us. It robs us of ⧗ **TIME** to think and reflect on what is happening to us.

I initially used this kind of conversation with my students as the first step toward getting them to engage in critical thinking, which has always been a central element to my teaching method. However, more recently, I have become concerned that it points to a deeper problem that they need to address, both for their sake and for the world that they will inherit. On a micro level, the world will not be kind to those students who cannot think critically. Rapid technological change will put a premium on those who can continuously adapt to their paradigms. This requires critical thinking and even more so the ability to engage in creative problem solving and constant redesign. They are used to being on rails with little control over their environment and only find themselves propelled forward by circumstance.

My students are poorly prepared for the uncertainties of the technological changes they will have to face over their lifetimes. The system gives them little time or incentive to reflect on what the purpose of their learning activity is. If we want people to be lifelong learners and innovators, we must get them to explore different ways of looking at the problems of learning and development. This requires them to find meaning in what they are doing. Locating meaning requires ⧖ **TIME** and we must learn to prioritize building ⧖ **TIME** into our educational and professional days for reflecting on what we are doing, where we are following outdated practices, and how we can take advantage of the new opportunities that technology is opening for us.

The common refrain is, however, the opposite. Instead of giving us ⧖ **TIME**, we often perceive technology as taking it away. We blame the always-on network for degrading lives, because it undermines hard-fought worker protections based on Industrial Age paradigms. (O'Carroll 2014) The internet and globally networked economy it made possible creates mechanisms for make-work such as Amazon's Mechanical Turk, which advertises itself as: "a crowdsourcing marketplace that makes it easier for individuals and businesses to outsource their processes and jobs to a distributed workforce who can perform these tasks virtually." (Mechanical Turk) This is nothing more than accelerated piecework on the old industrial model. Participants in Mechanical Turk, and other technologically enabled services such as Uber, Lyft, DoorDash, and Amazon warehouses, are essentially yoked to the machine in new ways. Instead of winning ⧖ **TIME** with our tools, we appear to have lost it. Instead of augmenting us, these new tools are used to take over our lives.

This is not the technology's fault. Technology accelerates culture. It does not change it. If the culture, which is a legacy of the late Industrial Age, demands finding the cheapest way to produce widgets, then the internet will happily facilitate with tools to do just that. If, however, we take the ⧖ **TIME** to redefine the paradigm, we may find that those perversities of the system fall away. This is yet another example of the old paradigm taking over the new with nothing more than a technological change of clothes. Instead of the conveyor belt, we now have the digital queue. This is not a recipe for creative thought or critical thinking. Instead of ⧖ **TIME** being spent rethinking what we should be doing, all of our ⧖ **TIME** is efficiently being taken up by doing what the old system compels us to do.

The Gig economy is not a recent phenomenon. It has antecedents that date back centuries. (Rab and Snell 1984) Industrial ⧖ **TIME** grew out of the cottage industries as a side effect of scaling individual work into collective work. It robbed us of agency by yoking us to machines that did not work when we didn't and required collective effort to function. The industrial ethos morphed ⧖ **TIME** into the scientific management of Taylorism, Fordism, and beyond. The 20th century saw this approach baked into the fiber of Industrial Age management approaches.

Prior to the Industrial Age, agrarian ⧖ **TIME** was linked to the natural movements of the earth and the seasons. Days were dictated by usable light and the usable heat generated by sunlight, which allowed our plants to grow, dictated years. Most humans adapted to this rhythm for millennia. This was never an easy life because nature can be a brutal mistress. It allowed for little ⧖ **TIME** for reflection for most. Technological efficiencies (or repressions such as slavery) only allowed a small elite to liberate itself from the constant drudgery of keeping food on the table. It was this elite that propelled us into the Enlightenment and, ultimately, the Industrial Age. Farmers became factory workers of the land as efficiencies allowed for fewer and fewer people to have to engage in the production of food. The rest of us moved to the factories where we became part of the technology of production, as described in the last chapter.

The legacies of these ⧖ **TIME** shifts are still with us today. We arbitrarily divide days into set periods of "work" and "leisure" that do not align with natural human rhythms. The week has arbitrary workdays and weekends that again have little function other than to provide us relief from the drudgery of the daily schedule we have imposed on ourselves. We gear everything toward production. This has not changed since we have moved from physical factories to virtual ones. Now our "product" is metric tons of unread reports instead of metric tons of steel, but the logic is all-too-often the same. Most of us are simply swallowed into the system and become either gears or product.

My students can't figure out their agency in the model of education because they are merely widgets in the educational system, and only accidentally its beneficiaries. ⧗ TIME is bent to produce widgets, whether those are made out of paper, plastic, or humans. It is difficult to imagine us solving the really tough problems of our time if we cannot resist this inertia and stop to reflect on where these conveyor belts are taking us.

This version of ⧗ TIME compartmentalizes thought. The problem is that thought doesn't work this way — particularly digital thought. Our brains don't work according to schedule. It is in those timeless periods such as walks, showers, and sitting on top of a cliff surveying the view that I find my imagination is most active. ⧗ TIME disappears when we engage in truly meaningful creative or emotional pursuits. It is in those moments that we throw out agendas, meeting times, class times, and deadlines and allow ourselves to connect ideas unexpectedly that true inspiration emerges. Entry into Csikszentmihalyi's *Flow* state is a timeless experience.

He states, "the sense of the duration of time is altered; hours pass by in minutes, and minutes can stretch out to seem like hours." (Csikszentmihalyi 1990, 2008, p. 49) Digital ⧖ TIME is free from the artificial constraints of industrial ⧖ TIME, which are dictated by extrinsic forces, not the intrinsic forces that drive creativity.

Industrial ⧖ TIME separates this kind of atemporal intrinsic creativity product from real "work" because it is hard to coordinate the means of production when not everyone is moving in lockstep. We often measure creative effort qualitatively. A "special place" is assigned to those who do it. Our organizations do not measure them in the same way as the vast majority of workers; they are considered extraordinary. Work today requires increasingly extra-ordinary efforts.

During the Industrial Age, the mechanism of ⧖ TIME was used to disconnect creativity from productivity. Widget time is fungible. If humans are widgets, we can swap them out like the machines that they serve. As Alexis McCrossen points out in *Marking Modern Times*, "For a number of reasons summed up in the words 'industrial production,' the value of labor came to be measured in increments of time, rather than by the tasks completed." (McCrossen 2016, Location 520) By decoupling task mastery from work and instead using the simplistic rubric of time on task, reflection ⧖ TIME came to be considered a luxury, vacation, or leisure, and fell outside of proper work. Innovative thought is merely a lucky byproduct of a system that measures productivity in this way. It also has the effect of decoupling work from what makes it humanly fulfilling. Machines are measured in uptime. Humans shouldn't be.

One problem with this logic is that we simply don't work on mechanistic schedules, at least not efficiently. Yet, we persist in thinking about ⧗ **TIME** in this manner. For instance, the schedule of my classes is deeply tied to the industrial culture of churning out widgets we call graduates. Work schedules operate on the same principle. We base both on the false transference of the idea of the shop floor that getting everyone to work in sync is the road to productivity.

My biggest enemy in any classroom is the clock, because learning doesn't happen on cue. I've seen workplaces time and time again cannot produce good ideas because everyone is on the same unpredictably manic track based on clock schedules. The "solution" many companies use to compensate for inefficient use of time is to extend the workday. Often this is doubling down on the failings that caused the inefficiency first place. This is a good way to achieve nothing but dead ends because no one stops to look around to consider alternative pathways to goals or rethinking those goals altogether using a different perspective. As Alan Kay once said, "point of view is worth 80 IQ points." (Kay 1984, p. 10) Digital tools give us the ability to change perspectives, but we have to allow ourselves the ⧗ **TIME** to do so. So far, most of us haven't done this.

Instead, adding a layer of digital technology onto this matrix has only made it worse. We assessed students and employees alike based on time on task as a proxy for measuring true output. Digital tools are used to gain ever more precise control over the daily, hourly, minutely tasks placed before both workers and learners. Communications networks are used to stretch these hours well outside the hours you have to keep the lights on in the office, resulting in serious quality-of-life degradations. Just as we discussed in Chapter 1.1, we are letting our tools for measuring and controlling ⌛ TIME enslave us rather than allowing them to liberate us. Aligning our tools with our humanity is the key to solving the challenges that they themselves have brought upon us. We just have to learn to slow down enough to think things through.

Digital technology alters our relationship with ⌛ TIME even more than it does with ✪ SPACE. It can make us very efficient or can provide a source of incredible wastage. Digital technologies decouple ⌛ TIME from presence. We are no longer limited in our productivity based on the hours we spend chained to a desk.

Our perception of ⌛ TIME is an artificial construct. The significance of this became apparent in the early part of the 20th Century as Albert Einstein's theory of relativity seemed to challenge even the scientific certainties of ⌛ TIME, touching off a debate between him and French philosopher Henri Bergson on just how far to take this. Einstein, sensitive about separating science from philosophy, rejected this mixing of disciplines (a feature of Industrial Age academia) and sought to maintain a clear delineation between the two. According to Bergson, "Time is for me that which is most real and necessary; it is the necessary condition of action: What am I saying? It is action itself." (Quoted in Canales, 2015, p. 36) Separating the notion of ⧗ **TIME** from clocks, the ultimate Industrial Age metric, is the critical factor here. While Einstein insisted on preserving the pristine nature of science, Bergson argued we cannot separate science from perception.

The temporal shifts created by the Digital Age should force us to reexamine Bergson's notions of ⧗ **TIME**. As long as we insist on binding ourselves to a Newtonian mechanistic notion of ⧗ **TIME**, we will miss many opportunities to find the critical meaning in our existences necessary for learning and innovation. The digital environment allows us to optimize our ⧗ **TIME** for work and gives us a great deal of ability to tailor our work periods around our maximum personal productivity. ⧗ **TIME** is a systemic construct. We should not underestimate how much this shift challenges our rigid perceptions of what is possible.

These rigid perceptions have channeled many of the gains made by our labors into greater efficiency instead of greater humanity. The last 70 years have seen unprecedented growth in productivity. According to the US Bureau of Labor Statistics, American workers are four to five times more productive today than they were in 1950. (Sprague 2014) If we translate efficiency into ⧗ **TIME** instead of growth, it can help us transition ourselves and our employees to the digital era, which will have a much more significant payoff in the long term. This "reflection" ⧖ **TIME** is essential for giving humans the space necessary to reimagine their challenges, creatively leveraging digital tools. If you do not stop and rethink what you are doing, the natural tendency is to continue as you are, even as that becomes increasingly out of sync with reality. Given how fast "reality" is changing these days, that strategy is no longer acceptable.

Industrial ⧖ **TIME** was rigid because it required the synchronous assemblage of large groups of workers to achieve productivity. McCrossen notes, "The far-reaching implications of this profoundly important system of social regulation, whose emergence depended on public clocks, pocket watches, and standard time, helped to define modern times." (McCrossen 2016, Location 248) We literally bent ⧖ **TIME** to accommodate the needs of the machines. The "regular" workday became the norm. Most companies still employ some format of the "9 to 5" schedule because it helps set expectations for meetings and collaboration, but this kind of shift work has its roots in the factory floor. This schedule can have pernicious effects and no longer makes sense because it is out of sync with the global rhythms of the digital age.

The global pandemic that struck in 2020 repeatedly showed how layering Industrial Age notions of ⧖ **TIME** and ✪ **SPACE** on top of a world suddenly forced into remote Digital Age spaces fails. Students forced to sit in Zoom sessions for hours and hours quickly burned out, tuned out, and lost what little meaning the Industrial Age educational system held for them. Workers stuck in endless slide deck-driven meetings had a similar reaction. What so many failed to perceive were the tremendous advantages these digital platforms and ⏳ **TIME** affordances offered. People could "assemble" for meetings from distributed locales. Conferences could bring in speakers otherwise unobtainable because of the costs of travel and lodging for them. I could integrate previously external support, such as a librarian, directly into the digital environment which my students operated in. Freed from the constraints of having to reserve physical spaces, I met with students in ways that made the most sense for them, not the building management. Digital Age businesses have been operating like this for years, with offices scattered across the globe. However, these businesses still represent a minority of the firms out there.

In a global communication environment, it is hard to imagine how most businesses keep "regular" schedules. If you align operations between New York, Los Angeles, Berlin, Tokyo, Beijing, New Delhi, etc., most of the world will seem out of sync. Digital capabilities now allow vastly disparate offices to work together increasingly seamlessly. Offices in China and India require 12 or more hours of temporal displacement from offices in the US. Good luck coordinating meetings that include Australia, the US and Europe. Digital communications have created vast opportunities to link offices, but the vast majority of offices still operate on industrial ⌛ **TIME**. This kind of decentralized synchronous work was very costly or simply not possible even a few decades ago. These affordances should change our perception of what an appropriate "work schedule" is in order to leverage talent and insight far beyond the boundaries of our local time zones.

There are still 24 hours in an Earth day, however, and one thing that decoupling ⌛ **TIME** from place in the digital era has spawned is a new awareness of human productivity cycles. Spanish and Islamic societies implicitly understand this better than the more synchronous cultures of Northern Europe. The human body goes through natural cycles during the day, and the time immediately following lunch is the least productive time for most. Daniel Pink calls this "the trough" and argues that the Spanish concept of "the siesta" tacitly acknowledges this reality. (Pink 2018, pp. 70-71) The industrial model of ⌛ **TIME** considered these kinds of notions backward because they didn't conform to the logic of the machine cycle. As we have seen, however, digital technology opens up the possibility of questioning temporal norms.

In a digital world, our schedules can become more human even as they become more global. The logic behind the industrial workday makes less and less sense the more interconnected we become. Digital is always on so that much work can continue any time of the day or night, with machines compensating for the vagaries of the human biological cycle. This is already happening as bots do most of the trading on international financial markets, taking over from humans to interact with markets that are out of sync with local time, to cite just one example.

Digital allows us to be more present through synchronous activities such as digital videoconferencing and meetings. It also allows us to work more effectively through asynchronous collaboration technologies, such as Slack or Google Docs. Modes of adaptive, persistent digital workflows, managed properly, also create opportunities for us to adapt our work schedules around more human pursuits that include ⌛ **TIME** spent with family and loved ones and maximizing our personal abilities by being sensitive to our biological clocks.

The pandemic revealed opportunities in many unexpected ways, as it forced humans to adapt to social isolation. Those of us that understood digital ⌛ **TIME** could adapt and grow as circumstances forced us to be socially distant. My classes are better than they were before the pandemic because of my ability to ⌛ **TIME** shift my instruction using a wide range of digital tools. (Haymes, 2020/2)

Like ⌛ **TIME** and ✦ **SPACE**, we often overlook the new possibilities that these realities are opening up for ourselves because we fail to slow down enough to examine the paradigms we operate under. Many companies explicitly reject the idea of toying with the 9 to 5 schedule even though they know full well that a lot of those hours are unproductive. Instead, they invest in digital tools to monitor employee's activities throughout the day, creating cultures of fear and surveillance that enforce conformity rather than the creativity that they claim to want.

These methods have bled into our education systems through badly designed learning analytics software that monitors the activities of students in similar ways in a vain attempt to "make" them learn better. Systems of education have developed poor ways of assessing student learning. Therefore, it should not come as a shock that learning analytics systems, which do little more than speed up measurement of those same variables, struggle to measure meaningful success. Several years ago, I was part of a team of researchers from the New Media Consortium that spent the better part of a year examining the potential of learning analytics. The almost universal conclusion was that the problem lay not in the technology, but in the lack of a clear consensus of what educational "success" even meant. This made developing a clear set of tools designed to measure and encourage desired outcomes an elusive goal.

Shifting these kinds of paradigms requires careful reflection and a willingness to examine deeply ingrained ideas and practices. This is only possible when we recognize the necessity to stop, step out of our daily cycles, and reflect on how we can adapt ourselves and what we are doing to the new realities of the Digital Age. This is what we should be buying with the increased productivity that technology is giving us; not simply doubling down on outdated paradigms.

Effective Digital Age ⌛ **TIME** management requires us to spend the asset of ⌛ **TIME** obliquely instead of in the headlong quest for productivity and profit. It requires organizations to live with unpredictable, open-ended outcomes, some of which will end up in dead ends, both personally and organizationally. Failure is deeply ingrained in the concept of ⌛ **TIME**. We will discuss this in greater detail when we explore the notion of "Play" in the next module of the book. How we think about ⌛ **TIME** is at least as critical as what we do with it.

One of the biggest conceptual mistakes that stems from the Industrial Age conceptions of ⌛ **TIME** is conflating ⌛ **TIME** with money. If "⌛ **TIME** is money" then that leaves little room for play. The Industrial Age lived and died based on Benjamin Franklin's homily: "Remember that time is money. He that idly loses five shillings' worth of time loses five shillings, and might as prudently throw five shillings into the sea."

Franklin's aphorism seems to transform ⌛ **TIME** into a fixed commodity that is analogous to money. We co-opted this aphorism over a century later to suggest exactly that. Franklin himself, however, apparently didn't see this maxim in the same way as later interpreters did. Walter Isaacson suggests that Franklin's, "goal was to help aspiring tradesmen to become more diligent, and thus more able to become able and virtuous citizens." (Isaacson 2004, p. 100) "Becoming able and virtuous citizens" is a qualitative measure, not the quantitative measure implied by money. In today's context, measuring ⏳ **TIME** *qualitatively* instead of *quantitatively* makes much more sense. We should endeavor to measure success through qualitative product, not time on task.

We must slow down in order to speed up. We must assess the quality of our ⏳ **TIME** and this may require ⏳ **TIME** doing nothing more than thinking about, researching, and discussing how we spend it. Unlike money, units of ⏳ **TIME** are qualitatively different from one another. The true skill of the Digital Age is using the tools available to us to assess how we maximize the *quality* of our ⏳ **TIME**. This is not something we were used to doing in an industrial environment where we expressed ⏳ **TIME** primarily in units of money, often in a direct sense. It also requires more sophisticated digital tools than the primitive counting tools that are used to measure productivity and learning currently. If we never demand these tools because we are still mentally stuck in the industrial paradigm, they will never emerge. This oversimplification is no longer adequate, and it has the pernicious side effect of facilitating our dehumanization.

Stepping out of this paradigm may be essential to our very survival as a species. It will certainly give anyone who does so a competitive edge in the short term. Humans are the most important part of cultural change. Cultural change is essential in allowing us to personally and organizationally adapt to the rapid changes becoming the norm in all industries. The most important thing we can buy with our ⏳ **TIME** is the ability to adjust and readjust to this fluid environment. We always have to recognize that human adaptation at this level is far from instantaneous. Institutions need to provide opportunities for humans to respond to change. ⧗ **TIME** is far less fungible than money and often requires a significant commitment of resources to take people away from their normal routine.

But a departure from routine is exactly what the Digital Age demands. "Routine" is something that machines do, not humans. One idea to maximize creativity in the face of rapid change would be to reward people adequately on the financial side but to tie rewards to reflection ⧗ **TIME** or time off. Reward ⧗ **TIME** with ⧗ **TIME**, not money.

There are tremendous multiplier effects to be had here. Studies have shown that time off not only has positive health effects but also enhances our productivity and creativity. (Seppälä, 2017) This should come as no surprise. Time off encourages employees to be truly productive in their time on task. Furthermore, "time off" allows them to reflect in a different context about the task or goals that they are trying to achieve. Reflection ⧗ **TIME** creates innovation opportunities for the employee to return to work invigorated with new ideas and motivations to speed up projects.

A Digital Age approach might be to reimagine the nature of "time off" altogether. As Daniel Pink quotes Steve Swasey, Netflix's Vice President for Communications in *The Flip Manifesto*, "Rules and policies and regulations and stipulations are innovation killers. People do their best work when they're unencumbered. If you're spending a lot of time accounting for the time you're spending, that's time you're not innovating." (Pink 2013, p. 24) Measuring ⌛ TIME as "work time" and "play time" is archaic and counterproductive. Creative work never ends and never begins. We must learn to blend both positively to be successful. This means making "work" less work-like as much as it means capturing productivity wherever it might occur.

As we explore in Chapter 2.3, the digital world allows us to do this in unprecedented ways. However, these affordances are unrealizable if we do not adjust our physical and temporal environments to allow this to happen. Looking back at the last chapter, we can see that ✪ SPACE (and its associated technology) cannot be separated from ⌛ TIME. In order to maximize the fragile, scarce resource that is ⌛ TIME, we have to move these fundamental considerations out of the way first. The physical and virtual have a direct connection to the temporal. Good design can optimize our ⌛ TIME. Bad design wastes it. We should not accept design that wastes our ⌛ TIME.

In the industrial era, when temporal considerations were primarily quantitative, these considerations were less important. There was already so much wasted ⌛ TIME in the typical worker's day. Poor design was simply noise. The implications of design were also less severe in simpler technological paradigms. Now, employees waste hours and even days trying to decode the latest poorly designed software platform installed on their computers and servers. The result is that we tolerate some truly dehumanizing working environments in the name of "upgrading" technology. The ⧗ **TIME** spent "working" when this happens becomes an exercise in unnecessary endurance, not one of quality work. Mindless application of technology is a "luxury" we can no longer afford.

Good design, however, needs to be complemented with mindfully carving out ⧗ **TIME** to engage in iteration and reflection. ⧗ **TIME** to fail and learn from failure is critical to grasping opportunities for innovation. Design thinking doesn't work if you put all of your effort into a bad prototype without spending the ⧗ **TIME** to evaluate its effectiveness critically.

In my field many educational innovations, from Learning Spaces, to MOOCs, to technological or pedagogical experiments such as flipped classrooms have failed to produce significant change because there was significant pressure to push forward with the "latest and greatest" without proper ⧗ **TIME** to reflect on their efficacy. Even the best learning or collaboration space design will fail if not properly evaluated and iterated upon. Non-spatial factors, such as creating opportunities for ideas to collide, facilitate the creative/innovative environment necessary for change. A mechanism for collecting, rigorously analyzing, and building on those ideas is absolutely critical to the long-term success and sustainability of these kinds of initiatives. This is a key insight of the Lean Startup movement in Silicon Valley.

Ideas need to be given temporal space to breathe. That involves finding ⧖ **TIME** to play with them. In the 1990s, psychologist Kevin Dunbar studied the origins of creativity and innovation in controlled studies of scientific laboratories. He came to the following conclusions:

> Conceptual change, like evolutionary change, is the result of tinkering. From a psychological point of view, this account of conceptual change explains why it is so hard to discover the underpinnings of creativity. The many incremental steps that are involved in creative cognition are often lost and forgotten, and the act of creation becomes a mythical entity in which the final step in the creative process is often seen as the cause of the new concept. This leads to the proposal of entities such as distant analogies and insight as more important in creativity than they really are. (Dunbar 1997, p. 488)

Dunbar posited that change came out of a constant reformulation of ideas. This is also a central point in Eric Ries's *Lean Startup* Model. (Ries 2011) Ries argues that successful startups in Silicon Valley do so as a process of trial-and-error and that "runways" (the time a company has to bring its product to market), should be measured in terms of time to "pivot" rather than money. The reason that the Lean Startup method is more workable in the digital era than it was in the industrial one is cost. The cost of failure is so much lower than it was when practically any product launch involved setting up a complex production facility to manufacture the idea. Nowadays, many of the most successful products are virtual and computer-controlled technology for producing prototypes has reduced the costs of product development even in the physical world. In larger-scale manufacturing, decentralized "plug-and-play" architectures have shrunk the size of initial investments. These factors all make "tinkering" cheaper. Ries argues we need to restructure our business activities around the one true scarcity: time. This is also true for individuals trying to navigate their own careers.

Technology is not the driver here. It is merely a facilitator and accelerator for rethinking processes. If behaviors do not reshape our behaviors, and this requires ⧖ **TIME**, we will usually fail to take advantage of most of the gains a particular technology offers. Further complicating this is that most technology vendors don't truly understand what drives productivity. Sure, faster and more powerful will always get you incremental gains. However, we can only achieve exponential gains by reinventing how we do things. A new accounting program will help with the former, but it cannot achieve the latter. Only ⧖ **TIME** for experimentation, failure, and iteration will allow the true innovation that leads to exponential change.

This is not a singular investment of ⧖ **TIME**. It has to become part of an ongoing process of reinvention. This is because innovation is nothing more than self-learning, scaled. Iteration is central to learning. Learning is core to facilitating individual innovation. What we traditionally think of as "innovation" is merely that individual innovation scaled up. Both require the strategic use of digital ⌛ **TIME** to maximize individual and collective opportunities for learning. If we do not give humans the ✲ **SPACE** and ⌛ **TIME** to achieve individual innovation, scaled innovation is not possible. All learners, whether of the formal or informal variety, need ⌛ **TIME** to process new inputs, share those half-formed ideas with others, and reconstruct their individual realities before reconstructing their community's realities.

If this ⌛ **TIME** is never given, nothing ever changes, innovation is stillborn, and the world moves on, gradually making an organization (and individual) irrelevant. We will explore fostering a narrative of change and adaptation in far greater detail in Chapter 2.1 as we look at how the new digital paradigms can facilitate creativity.

Fostering creative outcomes requires a willingness to accept open-ended outcomes. Once again, the new tyranny of digital measurement tools threatens programs that use ⌛ **TIME** in this way. "Quantifiable" accountability requires the setting of clear goals and benchmarks, the precise measure of milestones or waypoints, and adherence to rigid plans. We often see a questioning approach, particularly in a programmatic context, as a risk that is hard to quantify unless we provide some sort of predictable, often quantified, outcome. We perceive uncertain outcomes as not being scientific or data driven. Ann Pendleton-Jullian and John Seely Brown argue that today's problems can only be solved through a constant process of redesign. (Pendleton-Jullian and Brown 2018) They call this approach "emergent design."

> Making progress on complex problems that are not about things requires thinking and designing with an understanding that one cannot design for absolute outcomes. The future cannot be designed. The future emerges out of actions in the present as they are influenced and interpreted through actions of the past. One must design understanding principles of emergence. (Pendleton-Jullian and Brown 2018, Vol. 1, p. 44)

Pendleton-Jullian and Brown's approach is critical to approaching the challenges of what they term a "whitewater world" and apply to both individuals and organizations. They imply that all creative organizations have to reimagine themselves constantly. Ideally, we construct systems that force this to happen, for all innovation starts as a fragile flower. While we explore the role of scaling innovation through iterative play in more detail in Chapter 2.3, this kind of reimagining cannot happen without rethinking how we use our ⌛ TIME. Leaders have to learn to be mindful of the structures that are legacies of means of production that may no longer be relevant in an economy that increasingly rewards creativity and innovation over mass output of goods. Temporal reinvention applies to them as much as it does to any employee.

It's human nature to resist change. We are creatures of habit and constantly create structures to facilitate predictability. Creating ⌛ **TIME** for unpredictable activities has to be a conscious effort. Channeling these efforts back into the larger system is the last part of the IdeaSpaces framework and to this we now turn as we look at how environments form systems that can be designed to augment our capacity for growth rather than shielding us from change.

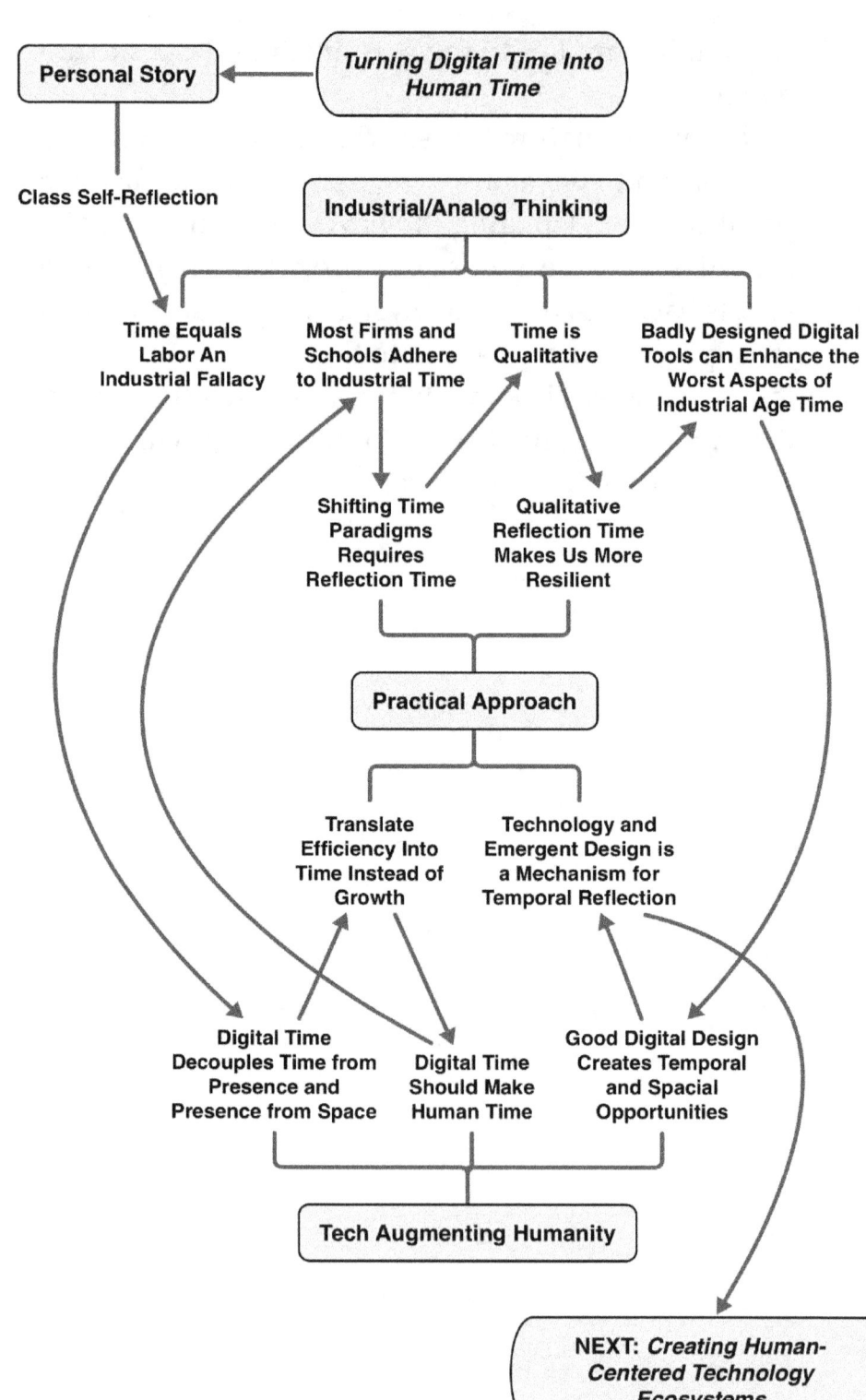

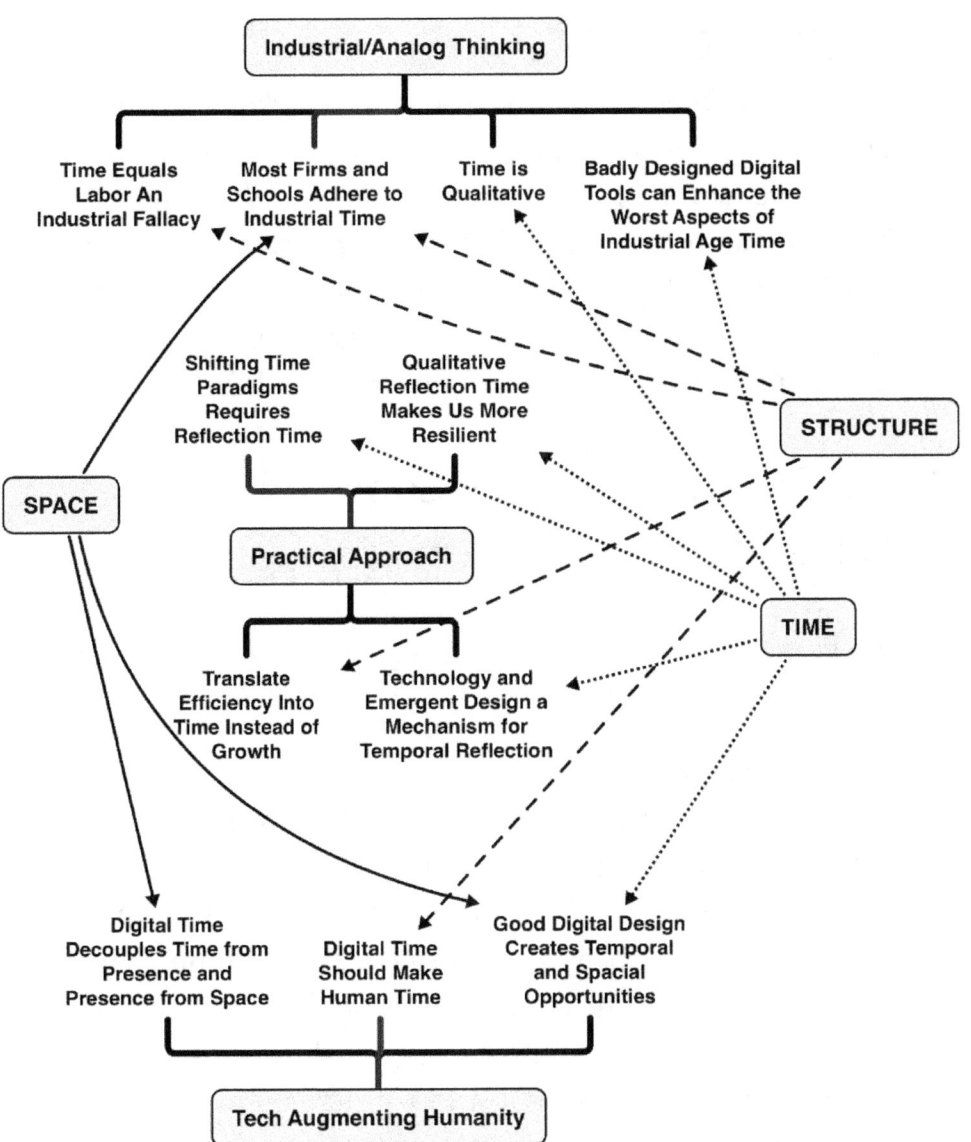

❄ STRUCTURE

Creating technology ecosystems requires thinking about technology as a structural element and expanding our traditional definition of technologies to include things like furniture, design, and architecture with the opportunities created by the Digital Age. Having a holistic view of your technology environment is a real accelerator for recognizing and seizing opportunities for innovation and learning.

1.3 Creating Human-Centered Technology Ecosystems

Function reforms form, perpetually.

Stewart Brand, 1994

In 2008, the Texas Legislature passed House Bill 2504. It stipulated that all public colleges and universities require all of their faculty to post their curriculum vitae, and the syllabi of every one of their courses for the previous three years online. This mandate carried with it no additional funding to set up the systems of support required to achieve its requirements, a not uncommon practice among state legislatures. Every public institution immediately began scrambling to find a solution to the new requirements. Along with Jordan Carswell, I led the team at Houston Community College in responding to this need. The system went online in late December 2010 after a little over six months in development and costing around $70,000. Within 6 weeks, over 4,000 faculty, with 10 minutes of training in video form, had uploaded the required documents. What was even more exciting was that they had also uploaded over 18,000 *non-required* documents to support their teaching mission. At a stroke, The Learning Web, as we called it, had reshaped the online environment that HCC faculty had to work with. They eagerly grasped at its technological opportunities.

What we had done was to create an environment conducive to supporting innovation and simply opened a door to our users. Outside of the Learning Management Systems (LMS), which housed online courses, there were few platforms in the early 2010s where faculty could easily store and share materials with their students. The Learning Web was an open system from the start. It had grown out of smaller efforts that the individual colleges of HCC had run on their own and had always been a faculty-driven technology. Jordan Carswell, currently Director of the West Houston Institute, created the original iterations using open-source content management software and adapted it to the specific functional needs that faculty had. In response, he created a very responsive workflow for facilitating their online needs of faculty.

When the Texas legislature issued its mandate, the instructional technology leadership at the college looked at the options and decided that scaling up the existing Learning Web efforts to create a system-wide ❋ **STRUCTURE** and to add easy file management for the required CV and syllabi. Jordan and I worked closely with a developer, Enfold Systems, to create a simple, user-friendly system that still offered substantial functionality. We resisted efforts from all sides toward feature-creep but built in an ongoing development budget so that faculty could continue to request additional features to support their teaching mission and to allow for performance upgrades as we saw the loads that the academic schedule placed on the system.

The Learning Web represents a ✱ **STRUCTURE**, in this case an electronic one, that took a basic requirement and turned it into a system that supported unexpected innovation. While this was always Jordan's and my plan, it was not really the priority of those who were funding the effort, who were mostly concerned with fulfilling the state requirement. To this day, the Learning Web remains one of the most popular and successful technological implementations that the college has ever implemented. It is also one of the most underappreciated because its functionality is so invisible to most eyes. The Learning Web just works and has had few technical issues. It is a ready set of tools to support any faculty member who wants to implement something online quickly. As new technologies, such as streaming videos, came online, we seamlessly integrated them into the functionality of the platform.

Structures like the Learning Web create the *opportunity* for innovation and learning. They do not create innovation and learning. However, they can certainly retard it. Structures come in all forms. We operate in a system of nested structural elements that range from the physical spaces that in which we live, work, and learn to virtual environments like the Learning Web or an LMS to organizational structures to the very communications networks we rely on to communicate with an ever-widening circle of friends and colleagues. Structures are systems of systems. They are human systems intertwingled[†] with technological systems. Structures are combinations of ✪ SPACE and ⧗ TIME extrapolated into systems that can scale innovation and learning. To achieve this, they must shape those spaces and the nature of the ⧗ TIME allowed for self-reflection. Structures are key to whether innovations can be scaled or not. They determine whether good ideas flourish or die in the cradle.

In the industrial era, these structures were necessarily rigid. Production of material goods required a high degree of collective effort and hierarchical management structures, sometimes extending throughout an entire state such as in the former Soviet Union. What became clear about these structures is that they overwhelmingly disempowered the individual and, as the collective subsumed the individual, innovation suffered. One of the fundamental problems with structures is that they are, by definition, rigid and resistant to change. Human

---

[†] "Intertwingled" is a term coined by Ted Nelson that means "To interconnect or interrelate in a deep and complex way." Source: **en.wiktionary.org/wiki/intertwingle**. We will explore this concept in more detail in future chapters.

thought is dynamic and fluid. Both are necessary for the augmentation of the human project (to borrow Douglas Engelbart's words), but we must achieve a necessary balance to nurture thought rather than stifle it. Johnson described the ideal condition here to be the establishment of structures based on "liquid networks." (Johnson 2010) Effective innovation and learning systems (or physical environments) will create opportunities for such networks to be formed.

As humans have transitioned to a digital existence, change and adaptability should become much easier. The industrial era programmed us to accept the output of production. With physical products, that toaster oven that came from the General Electric factory was what you got. Your selections were limited. With media, the contents of the network news, the bookstore, or library were what was widely available. Sure, you could write for yourself, but sharing required the intervention of a publisher and/or broadcaster. We similarly automated learning as schools and universities channeled knowledge. The Digital Age is changing all of that. Now, for better or worse, you can design and build your own toaster, create and disseminate your own media, and dis-intermediate your own learning.

The structures that we construct are the lagging feature in this equation. In 1994, Stewart Brand published *How Buildings Learn*, a book that every technologist, space planner, and architect should read. (Brand 1994) Brand is not an architect. He is a technologist, futurist, and writer. Here, his salient role is one of being an outsider. He approaches the study of buildings as technology systems. They are supposed to adapt and flex according to the needs of their human inhabitants. He criticizes spaces that don't do this as monuments to the architects' vanity. Most architects are completely unaware of this book because it isn't part of the standard architectural curriculum. As a result, they often continue to miss central points that Brand cogently picks out of building design, such as the critical connection between design flexibility and innovation. Crucially, environments should flex in response to the inevitably changing needs of their inhabitants.

Brand's vision is very much in keeping with our vision of technology as a servant of human activity. Buildings are structures and represent systems of technologies. They are crucial to the development of innovative and creative thinking that takes place within them. While Brand was writing before truly virtual environments such as Twitter, Facebook, or World of Warcraft, as a technologist he brought a digital sensibility to the study of architecture. Brand recognized early on that most systems of technology like our buildings quickly become invisible. In the words of Kevin Kelly, "Despite its power, technology has been invisible, hidden, and nameless." (Kelly 2011, p. 6) This invisibility masks the power these systems have over our lives.

This invisibility stems in part from an industrial paradigm that trained us to accept and take for granted static environments. I constantly encounter people who see the worlds they inhabit, both virtually and physically, as being fixed in nature. Brand challenges those habits of thought when he argues that "the most inventive creativity, especially youthful creativity, will be found in Low Road buildings taking full advantage of the license to try things." (Brand 1994, p. 24) The Digital Age augments our ability to adapt any paradigm and recombine its elements into more human-centered spaces that facilitate creativity, learning, and innovation. Any ✡ **SPACE**, physical as well as digital, can effectively become a "low road" environment. Achieving this requires us to adapt our conceptions of human structures and how they are influenced by the systems of technologies within them.

Structures themselves imply a larger purpose or intent is at work than the necessities of having specific tools available for whatever activity you are trying to undertake, whether that is in a physical, organizational, or digital environment. The ✡ **SPACE** has to project a purpose of human-centeredness and even the most innovative spaces can fail because, in the headlong quest for novelty, they overlook the humanity that they are trying to serve.

Those designing environments must never lose sight of the fact that they all tell stories. Some stories encourage human innovation and creativity. Others stifle it. Human beings are storytellers, who will instinctively construct mental models of any environment they inhabit. This habit is fundamental to what makes us human. Those of us who could detect patterns had an evolutionary advantage. Patterns of jungle disturbance might indicate the presence of a tiger or show us where to find our next meal. If early humans failed to recognize the "story" the jungle was telling them, they were liable to become lunch or starve to death. Those who could convey those stories to their offspring made sure that they also survived at a higher rate, and so on. Now we construct those patterns ourselves. Structures tell stories and a lack of attention to this critical factor often causes them to fail in their intended purpose.

Stories in the Digital Age can be profoundly disorienting and ignoring this central reality can determine whether a particular environment succeeds or fails. The old world was scissors and paper. This new world is cut-and-paste. Cutting-and-pasting text is not a new thing, but, until the digital age, we could not undo or redo that first cut. These linear, textual structures of ideas that we organize into sentences, paragraphs, chapters, and books form the basis for one kind of story. This book is a constructed mental environment and, if I have done it correctly, you are as fully into the narrative here as if we were sitting in a room interacting with each other.

The Digital Age opens up a whole new range of possibilities for telling stories, which we will explore much more fully in the next chapter, but in this context, we need to be sensitive to the fact that these are all constructed environments. While Brand takes a functional approach to his technology environments as technologists, we must understand how profoundly disorienting a world where everything is malleable is to modern humans. We instinctively make up stories to explain any situation that confronts us, but now those stories can change faster than ever before. Narrative infuses everything we do and, as designers and users, we must be sensitive to the stories our technologies tell, whether those are an iPhone or an auditorium or a Human Resources department. Constellations of technology and physical spaces form narratives every bit as powerful as constellations of words and letters.

Text and letters themselves formed the basis for libraries and universities as ideas intersected to create new and powerful constellations of intellectual energy that spawned the modern age and what we understand as technology itself. The university is itself a technology for creating intersections of ideas and, by extension, technologies that are created by those ideas. Increasingly, we can say the same for businesses, which traffic in ideas far more these days than they traffic in actual goods. The university has long been a tool making tool. As Steven Johnson points out in *Where Good Ideas Come From*, innovations often come from the unexpected intersections of technology and ideas. He writes, "innovative environments are better at helping their inhabitants explore the adjacent possible, because they expose a wide and diverse sample of spare parts – mechanical and conceptual – and they encourage novel ways of recombining those parts." (Johnson 2010, p. 41)

I would take this one step further and point out that creativity and innovation occur when technologies fade into the background. This is because we build technologies on top of other technologies. We often take the end of the underlying technology for granted when it is a product of an earlier innovation. We don't notice most underlying technologies. They fade into the environment. We only notice roads when they fail and become riddled with potholes, to cite just one example. We stop considering many technologies because they become part of the systemic pattern.

This is as it should be. Technologies have to become invisible before we can use them for true creative activity. We don't think of the university as a tool, much less a technology. It's a collecting place for people and ideas. As we discussed, the master carpenter doesn't think about his tools. He thinks about what he's building. In much the same way, we are unreflective of the spaces we inhabit until they inhibit our ability to get our work done.

As we create collective and collaborative workspaces, we have to create systems of tools that become invisible so that we can create within them. These spaces also need to tell stories of inclusion, empowerment, and general play so that the humans that inhabit them instinctively realize that is why they exist. Achieving this kind of story means that we finally have to realize the goal of ubiquitous creation technologies. The ease of use of the Learning Web pushed the technology into the background and it was this fundamental feature that led to its rapid incorporation into the workflow of thousands of faculty members. The Learning Web was fully virtual. This was novel and what was possible in the early 2010s. Now, it is much easier to create environments that can seamlessly span the virtual and physical worlds if we look for those opportunities.

Good structural toolsets need to sit in the background, ready for us to grasp them when we need them. We need to tailor them to use the strengths of a particular technology, such as the ability to store persistent data and to rearrange it in a fluid environment to create, capture, and share innovation. Much like the author and editor, we need to figure out how to tell stories with our spaces that encourage creative human activity. We already have the basic infrastructure of technologies. It's merely a question of putting these pieces together to create ubiquitous and seamless technology platforms that augment human intellect.

Awareness of non-traditional narratives such as those expressed by our technology and physical spaces is critical in understanding how they impact the people that inhabit them. This is because the basic storytelling instinct follows us everywhere we go, even extending to our perceptions of the physical spaces in which we work, teach, and learn. It's not just the speaker that impacts the story being told in a room. It's also the ✿ **SPACE** itself. All stories are subject to misinterpretation. The story received is much more important than the intended telling.

To complicate this picture further, not everyone is telling the same story. Architects are telling a design story which may or may not conform to the end users' stories. Planner's stories are likely to revolve around budget, standardization, maintenance, etc. and these may often, especially in the middle of an implementation of a large-scale ✪ SPACE project, take precedence over the stories that facilitate effective creativity. Builder's stories revolve around completion deadlines and the myriad disruptions that can occur along the way. Since the industrial system rewards quantity over quality, these kinds of stories can be most destructive, as the successful contractor is one who completes one job after another, regardless of whether the result serves the goal. Developers are digital contractors and operate in much the same way and can have similarly pernicious effects on the final product if the end goals are not at the forefront of all activities. These ❋ STRUCTURES, even as they construct structures themselves, are subject to their own internal logic.

Conflicting rigid narratives create conflicting rigid tools. The most difficult element of any construction or software development process is keeping the ends of the ultimate inhabitants/users of the ✪ SPACE at the table throughout the process. There is no excuse for not doing this in a Digital Age where information can flow between silos to develop new understandings and fused narratives. There were many good technological ideas placed before us during the development process of the Learning Web in 2010 by the software developer. However, we tabled many of them because they did not contribute to the needs of the users. While they would have added additional capability, they did so at the price of complexity. The key to creating the best possible environments is getting these groups to maintain constant communication with each other and a representative of the end users throughout the design and implementation process. Having a connector who understands the priorities of both groups and mediate processes is absolutely critical in this process. I performed a similar role in both the development of the Learning Web and the West Houston Institute (see Chapter 3.2). Preserving the design vision, informed by a process centered on user needs, is essential in ensuring that the story told in the ✿ SPACE conforms as closely as possible to the mission of innovation and human augmentation that we all wish to play out there.

Buildings and rooms are physical manifestations of mental structures. In order to have a human-centered ✿ SPACE, it has to tell a human-centered story. It's hard enough to change cultures without physical barriers compounding the difficulty. This is most easily illustrated by showing physical spaces. However, the same principles apply to virtual and organizational spaces. In all cases, ❀ STRUCTURES that are isolating, hard to understand (such as bureaucratic mazes), and inflexible to the needs of the people they are supposed to serve are profoundly disempowering. They stifle innovation and human augmentation. Understanding the stories a technological ✡ **SPACE** tells us has a major impact on whether the ✡ **SPACE** is maximally effective. We have to learn to ask questions like:

- What kinds of messages are being sent by our structures and spaces?
- Are they encouraging exploration or conformity?
- Are they encouraging inclusion or hierarchy?
- Are they encouraging creativity or passivity?
- How effectively is creativity being distributed?
- How are they affecting the ability of a speaker to tell his or her story?

Let's look at the stories several kinds of physical spaces are telling us.

Conference Room

This ✪ **SPACE** at first glance is a collaboration ✪ **SPACE,** but it still has a front and back as defined by the screen. By using the glass as a work surface, it is possible to change the story of this room somewhat, but its default position is to emphasize a unity of direction toward the large screen in the front.

A Unidirectional ✪ *SPACE*

It still amazes me how many classrooms and meeting rooms look like some form of an auditorium. Rooms that have a front and back imply asymmetrical power structures. Asymmetrical power structures, like those in the Soviet Union, stifle innovation and ideas and make organizations inflexible in the face of change, whether that is technological, social, or economic. Often, the cost of technology dictated its rarity. A school or business can only afford one screen (be it TV or projection) per room, so that naturally becomes a focal point. Anyone standing in front of it becomes a de facto leader of the group. Even when the technology becomes cheaper or the institution in question has a significant increase in budget, this budgetary limitation drives standards into the future. Instead of completely rethinking the technology in a shifted paradigm, the old paradigm is just replicated endlessly. You almost always end up with a scenario where quantity trumps quality. Innovative systems invest surpluses into expanding creative space, whatever for that might take.

Unidirectional types of rooms have their purposes, but they are at odds with collaboration. In these kinds of spaces, information flow is heavily weighted in one direction: from the leader to the followers. The right leader can work against this to some extent, but he or she is essentially working against the ✿ **SPACE** instead of within it. Even the most elegantly designed collaboration ✿ **SPACE** will only serve its purpose if the group using it knows how to take advantage of its qualities. It helps when it tells a story of collaboration, but the "environment" comprises far more than physical systems. A cultural system that encourages exploration must complement any innovation project. The goal is innovation, not the spaces themselves. It's easy to show off shiny technology. It's a lot harder to show the impact it has on the culture over time. Often, we neglect this outcome. We end up bringing "innovative" spaces into conformity with the existing cultural paradigms. The results are predictably disappointing.

Active Learning Classroom

Active Learning Classroom in Use

We deliberately designed this ✪ SPACE to flatten the environment in the room. One approach to forcing cultural change is to create spaces that are physically inflexible and force the story in certain directions. The Active Learning Classroom tells the story that the inhabitants (in this case students) are going to be the center of activity within the room as opposed to one inhabitant dictating everything that goes on there. The monitors represent windows to the world and, since the students working in the pods control them, they are empowering work environments that leverage Digital Age technologies.

The circular nature of the tables, coupled with student whiteboards that create a collaborative workspace, conveys the possibilities of the ✪ SPACE. This image shows a blending of physical activity (students hacking problems as a physical group) and shared virtual activity (the use of online concept-mapping software — discussed in Chapter 2.1). These ✪ SPACES constrain the physical environment to benefit the virtual one. However, the physical environment and virtual environments should complement each other, leveraging the strengths of both paradigms. Crossing the digital boundary in this space depended entirely on students being able to connect easily and on demand. The technology of the 2010s struggled with this requirement. When using this room, I often defaulted to the student whiteboards because getting even one student per table connected to the digital displays consumed time and effort, interrupting the flow of teaching and learning. In the

Digital Age, we are going to see an increasingly blurred relationship between the physical and virtual. The design of both environments should complement each other to maximize the opportunities offered to us by our Digital Age tools.

Just because a ✿ SPACE is virtual doesn't mean that can't suffer from similar structural issues as physical environments. Learning Management Systems have been around since the 1990s. They grew out of the modalities of Usenet discussion groups, databases, and email. Mechanisms for evaluating student work and roster management (another form of database activity) were added to these tools and the "modern" LMS was born. Suddenly we could scale learning out to a much larger audience unable, because of distance or ⧖ TIME, to attend classes in a more traditional manner. As a result, schools experienced rapid enrollment growth in their new online sectors. What the designers of Learning Management Systems overlooked was the nature of the environments they created. At a time when we saw increasing calls for active and empowered learning, the LMS effectively enshrined the lecture method of instruction, because that was all it could do with the tools it could offer. The LMS is much like an auditorium classroom. Information flows from the front to the back. To this day, most people use Learning Management Systems primarily as content management systems, places to store and organize faculty content. If all you are delivering is content, you are creating an environment very much in keeping with outdated industrial era mechanisms of teaching, which primarily featured the delivery of content by an "expert" standing at the front of the room. Only now, there is no aspect of human contact. Even the Socratic method of teaching is virtually impossible. As a result, we have created profoundly disempowering online structures that were driven by the original development of technologies available 25 years ago. Like the expensive TV at the front of the room, the story of the ✿ SPACE has persisted long after its technological or budgetary rationale had passed. Learner-centered design has never been part of this process.

Most businesses lag in online environments directed at their internal audiences. This is because they can compel their user bases to endure atrocious experiences. Some of the worst environments I've ever seen have been corporate "training" platforms. The goal of these systems is compliance. Learning outcomes are commensurately low.

The Pandemic of 2020/21 showed us what was possible when we exposed these systems to extreme stress. The best companies and institutions of learning created new environments instead of mimicking old ones. Failures were often ones of imagination rather than technology (although the economic circumstances of impacted students and workers exposed huge inequities in access). The pandemic should have been a stimulus for innovative practice and finally grasping at tools that had been at our disposal for years. This was not always the case.

Failure to appreciate and incorporate a deeply mindful approach to designing spaces, virtual and physical alike, can make the innovator's job that much more difficult. The two hardest things I've done in my various careers have been teaching and fostering innovation. Both things have one thing in common that sets them apart from the rest of my activities: they both involve stimulating curiosity in others. One of the enduring truths of human behavior is that curiosity is the most powerful motivator for creation and exploration. We want our students to explore and create new meanings as a central part of their college education. We want our faculty and staff to create novel ways to get them there. We want our employees to explore new ways of doing things in a rapidly changing world in much the same way. Our structures, physical, organizational, or virtual, either support this kind of activity or they don't.

My experience of creating fluid environments and looking for points of rigidity in my practice and approach served me well when the pandemic forced me to improvise. Remote teaching, presentations, workshops, and meetings gave me the opportunity to create whole new ❋ **STRUCTURES** based on the tools available to me. I could remix synchronous and asynchronous activity in novel ways to create environments for those that I was trying to reach. I detailed this reinvention of my teaching strategies in my book, *Learn at Your Own Risk*. (Haymes 2020/2)

Our ✪ **SPACES**, both physical and virtual, tell stories that affect how we communicate with one another. Our technological ecosystems tell implicit stories and create conformities among the humans that operate within them. We must carefully weigh notions of conformity, "front" and "back," flexibility, and adaptability as we approach digital design, whether they are on a physical, organizational, or virtual level. ✪ **SPACES** can push the intersectional conversations or retard them.

The design process that creates these ✪ **SPACES** themselves must reflect this kind of open dynamic. Digital structures allow humans to write their own stories, whatever those may be. This means that the design process itself must be as inclusive as possible in order to create structures that are themselves inclusive. As Melvin Conway wrote:

> [O]rganizations which design systems (in the broad sense used here) are constrained to produce designs which are copies of the communication structures of these organizations. We have seen that this fact has important implications for the management of system design. Primarily, we have found a criterion for the structuring of design organizations: a design effort should be organized according to the need for communication. (Conway 1968, p. 31)

Commercial Makerspace

Makerspaces are a physical manifestation of storytelling devices. Done correctly, they tell a story of ✿ **SPACE** (and its associated technologies) coming together in service to the creative hacker. Makerspaces are task-driven and unpredictable. Their design and management must reflect this purpose, or they will not work as intended. Inhabitants generate their own solutions to the challenges put before them. This requires a ✿ **SPACE** that is easily reconfigurable, with a wide range of tools and technologies available to them.

Makerspaces and other Creation Spaces offer unique creation opportunities for their inhabitants. The furniture is again very mobile (although somewhat constrained by the technology on them) and the floor design encourages a wide range of activities. The narrative here is one of functionality in the service of the human inhabitants' need to create. This should be clear to anyone watching the activity within the ✿ **SPACE** or taking part in its activities. Within the bounds of safety, the management ❉ **STRUCTURE** must support a certain level of unpredictability, or the discontinuities of the two stories will inevitably lead to failure. You cannot ignore the multiple layers of structures necessary to achieve the mission of the ✿ **SPACE**.

All communal spaces should employ the Makerspace ethos. We are always "making," whether we're talking about ideas, concepts, projects, or physical products. Stewart Brand would call most Makerspaces the epitome of his "low road" environments. These are spaces that humans can reconfigure infinitely to suit the needs of the inhabitants. In Brand's examples, they were spaces that had little aesthetic value and therefore could be demolished and rebuilt at will. One example both he and Steven Johnson cite is MIT's legendary Building 20. This was a temporary building erected during World War II that survived into the 1990s. It was the cradle of everything from linguistics to communications science to computers and the internet. (Brand 1994, pp. 24-28) This building was repeatedly reconstructed to fit the needs of its inhabitants because its story was always one of incompleteness and reinvention. The stories that drove it were those of its inhabitants. They were not driven by some unseen designer trying to predict an unknowable future. At their best, Makerspaces are created with this kind of attitude in mind. We should build them around stories of possibility, not of constraint.

Stories are at the heart of everything we do, but the surrounding structures always constrain them. Technology increasingly allows us to create ❋ **STRUCTURES** that facilitate innovate storytelling. Digital Age toolsets provide the opportunity for us to weave these stories seamlessly through virtual and physical ✪ **SPACES**. Everyone involved in the design process, from designers, to implementers, to administrators, to teachers need to be mindful of the importance of the stories they are creating as they contemplate the shape of any physical ✪ **SPACE** because those are the hardest to change if designed incorrectly. Innovative structures can turbocharge the stories of creative innovators of all kinds. If our goal is enabling humans to innovate or learn, we must nurture these stories wherever possible.

Digital affordances have given us unprecedented control over the structures we can imagine. A telephone existed for one reason. It was there for sending voice signals from Point A to Point B. "Form follows function" certainly applied to these kinds of analog technologies. The shape of a Walkman was driven by the cassette tape in it. The shape of the cassette was driven by the magnetic tape that formed its core. In the Industrial Age, the functionality of the underlying technologies often dictated form. Industrial approaches to designing systems and the stories that they tell are increasingly anachronistic.

The Digital Age represents a world that is vastly more customizable and subject to rapid iteration than ever before in human experience.

Digital Age design explodes these constraints and, over the last decade, has increasingly encroached on tangible, physical technologies from telephones to cars to buildings themselves. The implications of this will be profound. When everything is reducible to zeros and ones, we can change the shape of anything. The ripples of this new reality extend far beyond computers or even the iPods that replaced the cassette and CD-Walkmans. Those were just the first steps.

These shifts have implications for architecture, education, and work. Chalkboards (or whiteboards or displays) define classrooms. Their scarcity drives the lack of classroom space. The shortage of "instructional" space drives the need for restrictive scheduling that corresponds to its own internal logic, not that of learning. Learning, hard enough in an unconstrained environment, is now required to happen according to an arbitrary schedule. The logistical need to segregate also feeds specialization in the academic environment. From 10 am to 11 am you are in History class, not English class. (Haymes 2020/2)

This is also true of the world of work. We are already seeing much more fluid work environments emerging. My wife's insurance company recently downsized their ✪ **SPACE** and most of the employees spend 40-60% of their ⧖ **TIME** working from home offices. They work in digital environments, so the cost of admission is a laptop connected to the internet. The workforce is spread over several continents, so frequent physical meetings are consequently not practical. My consulting business suffers from similar constraints. Modern video conferencing technology has become so simple that a $15 subscription to Zoom is all that I need to be always on and available to clients and for impromptu meetings. I have turned these environments into a global Building 20 where I can meet on the fly and build virtual analogues of formerly physical environments constrained by ✪ **SPACE** and ⧖ **TIME**.

This level of adaptability became essential during the pandemic. Most digital environments are not as intimate as physical interaction in a room. However, they opened a whole new range of possibilities for human interaction that I often overlooked before. Meeting with students individually became far simpler when the overhead was nothing more than firing up a Zoom session. No one had to take time out of their day to drive to a campus at a set time. It was simple to pull up the student's work through screen sharing and work through problems together.

We still bind ourselves structurally to old cultural and industrial paradigms of putting workers into physical spaces together and imposing arbitrary sets of ⧖ **TIME** on them ("working 9 to 5"). However, I am writing this at 10 am on a Sunday morning in my living room. I have a full slate of work related to teaching and consulting responsibilities ahead of me, some of which I will get done today and some of which will slide to tomorrow. My narrative of work is that of managing a river of information and production. It's not that of punctuated states of ⧗ **TIME** where I have to check into a physical ✪ **SPACE** (the only exception to that is teaching, which is essentially a series of regularly scheduled meetings with my students). My narrative has been subtly shifting in this manner for years.

We need to build our physical, virtual, and organizational ✺ **STRUCTURES** with these new realities in mind. If the ✺ **STRUCTURES** we exist in dehumanize us, we need to get to work hacking them like we would do any other technology. Working, teaching and learning are likely to change considerably over the lifetime of the structures we are now creating and so those structures, while necessary, need to be as fluid as possible.

Technology is easier and easier to adapt to human needs. We must carefully evaluate any barriers that we encounter to see if it is truly a technological barrier or whether it is really our ✺ **STRUCTURES** that are holding us back. ✺ **STRUCTURES**, whether those are physical, organizational or virtual, are the most rigid parts of our technology systems. They are a necessary part of our existence, but we should always be mindful of how they limit what we can do. We need to recognize that structural constraints are likely to be undermined at an exponential rate going forward.

Our guiding principles should always begin and end with the human in the equation. Fluid, Digital Age technology is giving people a vast range of choices. This can become overwhelming, especially to those with many other decisions to make in their day-to-day existence. The strategy behind mastering these new realities is to focus on the intent of the activity and to do so critically. As we were building the Learning Web or the West Houston Institute, we never lost sight of the needs of the humans who would inhabit those spaces.

While the technological means may be constantly shifting, the ends rarely change much. We want learners to emerge from schooling with the knowledge, skill sets, and mindsets that will help them succeed in life. We want workers who move the company forward and leave with the satisfaction that they have contributed to the collective effort. Let's put that up front and build our structures backwards from there. We have the tools to do it and the Digital Age is making those tools more accessible all the time.

Even as we appreciate the power of storytelling and its relationship to the physical, virtual, and organizational structures, we also have to recognize that stories themselves are being bent in unprecedented ways. This area is most impacted by digital technologies, as the informational spaces, speed, and structures have been profoundly upended by ubiquitous informational flows. How this has bent our narratives is the subject of the next section of the book.

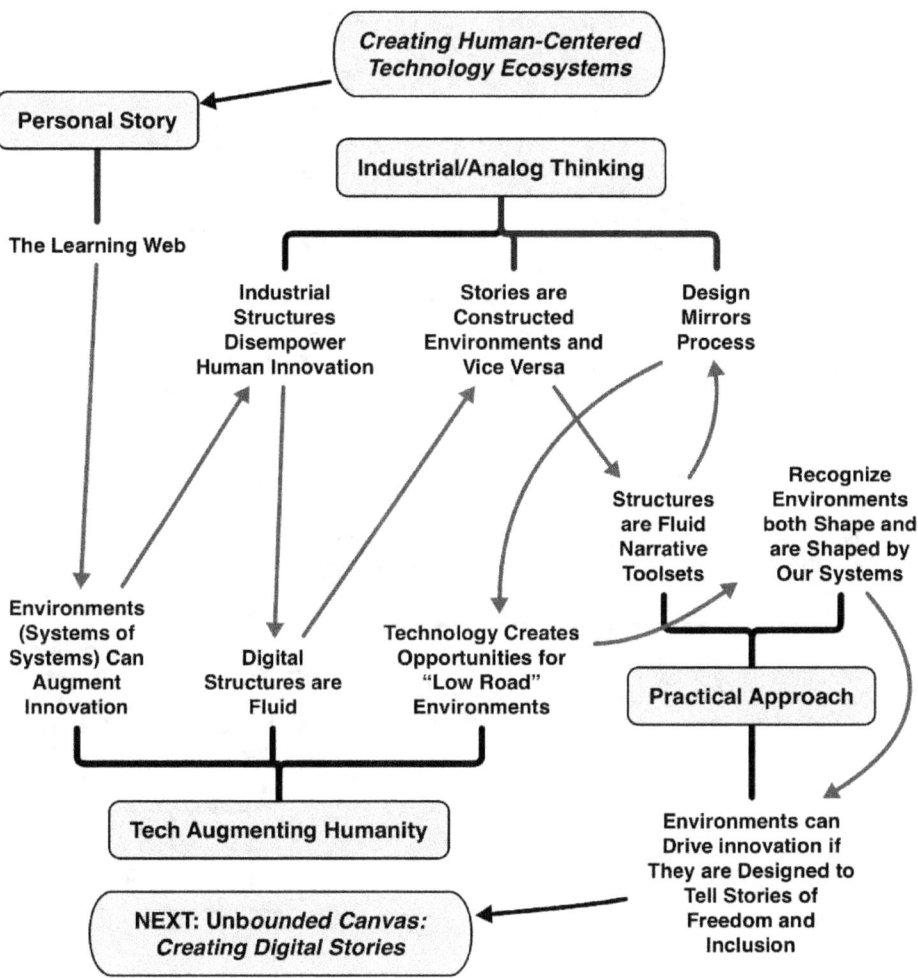

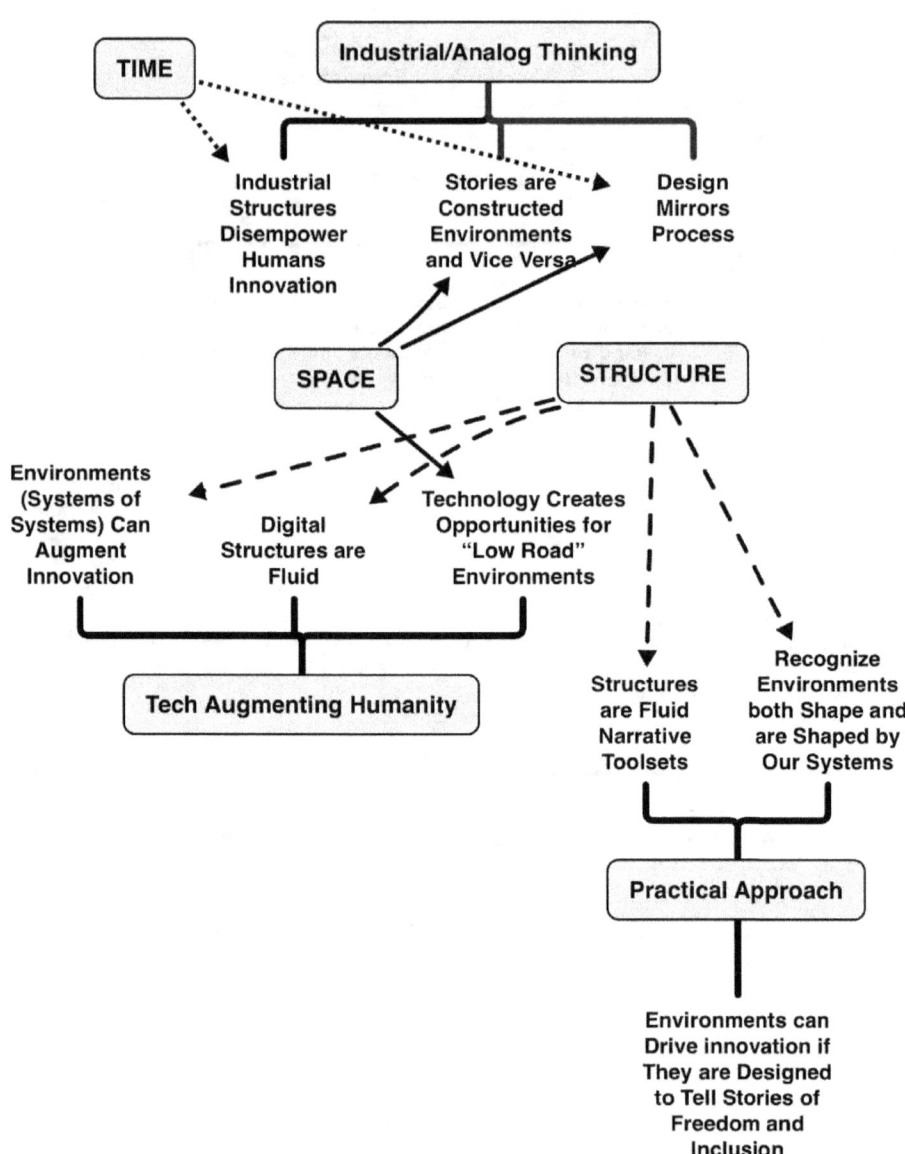

2.0 Creating, Mediating and Advancing the Human Story

Meanwhile, the poor Babel fish, by effectively removing all barriers to communication between different races and cultures, has caused more and bloodier wars than anything else in the history of creation.

Douglas Adams

✲ SPACE

Narrative ✲ SPACE is every bit as important as physical and technological spaces. Indeed, through their design (or lack thereof), technologies and physical spaces also tell stories. Understanding the new opportunities for communication and expression in the Digital Age allows us to create mental spaces that are conducive to innovation and learning.

2.1 Unbounded Canvas: Creating Digital Stories

An intellectual is a man who says a simple thing in a difficult way; an artist is a man who says a difficult thing in a simple way.
Charles Bukowski, 1969

Creator and reader are partners in the invisible, creating something out of nothing, time and time again.
Scott McCloud, 1993

I was an early adopter of word processing. As I related in the Introduction, when I was a freshman in high school, my father bought our first computer. Why this is significant is that it came at a pivotal point in my writing career. It was about this time that I was first forced to write long form prose and fiction. Because of this fortuitous technological shift, I learned to write on the infinite canvas of Word Handler, not the regimented canvas of the IBM Selectric that had graced our household during my early school years. I have always written digitally. My writing style is more akin to editing than traditional forms of writing. I worry about getting the ideas out of my head and into "the machine" first and then worry about how to organize them later. The ability to take digital bits of detail and rearrange them in any way I wanted to sort them has always been a routine matter to me. It was years before I realized how different this made me from my peers, much less from my teachers. I viewed text, not as a stream of information, but as discrete chunks of narrative that I could move around and place in any order that I needed them. The narrative structures that knitted them together could be added later. Instead of working on lines of text, I have always seen myself working on a canvas of information.

My formative experiences with writing and technology shaped my view of the storytelling structures under which we exist and contributed to my adoption of a metanarrative approach to all communication. I view communication as separate from the tools we employ to transmit and share ideas. Tools augment or impair our ability to create, transmit, and receive information, but they do not represent the underlying idea of creation itself. That would be like saying a piano makes Beethoven.

Text is only one point of departure, particularly in the Digital Age. Photography has stimulated a fascination with visual narrative in my mind, but it is not the camera that does so, it's working with the limitations inherent in taking the viewer to a time and place that is ephemeral in my mind's eye. Finally, as a social scientist, I am an observer of the storytelling of others. Media is used to channel large groups of people into constructing firmly held narrative beliefs. This landscape has become vastly more complex in the Digital Age and ever more sophisticated visualizations appear every day on YouTube, Twitter and Facebook. Understanding how to construct them, and how they are trying to influence the viewer, is critical in navigating today's digital information environment.

Digital technology is deeply empowering because it allows more and more of us to construct our own media worlds. My early adoption of word processing taught me I could boil everything down to digital bits that could I could reshape and manipulate at will. As Matthew Kirschenbaum put it in his excellent exploration of the history of word processing, "Rather than a reimplementation or remediation of typewriting, I prefer to think of [word processing] as an ongoing renegotiation of what the act of writing means." (Kirschenbaum 2016, p. 23) While writing continues to be a central tool in my arsenal of communication, I have learned to use conceptual canvases far more broadly in pursuit of narratives that serve my readers and students.

My early exposure to digital freedom liberated me, but it has not come without its drawbacks. One problem I have as a writer (and which I have struggled mightily with in this book) is that learning to write in an undisciplined environment may allow for freer exploration of ideas but having the touchstone of more restricted writing methods is sometimes critical for creating coherent longer narratives. Kirschenbaum documents many instances of writers even today relying on archaic methods such as typewriting or even writing by hand with fountain pens in clear opposition to the technological winds.

Consumers of information produced in this manner are even more profoundly tossed on waves of information. Navigating this boundless environment can be profoundly challenging and disorienting. Doing this on an organizational or societal basis is even harder. We will explore humanizing the consumption of information in the Digital Age in more detail in the next chapter.

In the IdeaSpaces framework, narrative technologies form the spaces for ideas. They may appear to be intangible spaces, but spaces of the mind are often as rigid as the rooms we discussed in previous chapters. Like physical spaces, their parameters dictate how those ideas are presented, communicated, and analyzed. Spaces of mind and imagination shape the way we can adapt those ideas, combine them with the ideas of others, and address the complex problems we face today. The ideas we can express in a poem differ from those we can express in a photograph. The medium profoundly shapes the message.

Like physical structures, the technology we adopt constrains how we process and communicate ideas. Narrative itself is a technology, but technology in the more traditional sense of the term is also a handmaid to our storytelling aspirations. Technical leaps in our narrative technologies such as the invention of writing, the mass production of writing, and the expansion of our vocabularies into the visual realm shaped how we create and relate our stories to one another.

For most of human existence, our narrative technologies were limited to oral communication, with some limited visual cues. Writing and literacy opened the text to an elite trained in the tools to consume and create in that medium. Industrial technologies in the 19th Century allowed us to scale literacy on a vast scale, but this created novel problems as knowledge outgrew our individual capacities to process it.

The tools we had to express ourselves, ranging from the machine press to broadcast television, had the effect of creating an elite of editors that shaped narrative according to their own preferences, some good, some bad. This also impacted how we learn as our structures of we shaped learning into increasingly specialized silos of information, each with its own set of gatekeepers and editors. Industrial narrative therefore assumed a linear nature in order to accommodate the technical ("sources" to "readers") and the resultant systemic constraints of the mass production of information and knowledge.

Technology has always provided sets of tools that can augment our ability to overcome our human limitations, leaving our cognitive ability to shape ideas as the primary limiter for creating narratives. We have adapted our cognitive abilities to suit the limitations of whatever technology we are using. Industrial Age technology shaped our thinking along linear, analog tracks. Technology has always been necessary to create most forms of the music and art (also forms of communication) that are central to our most profound human experiences. It still shapes our narrative possibilities in the digital world, but those constraints have changed as the tools of creation and distribution are democratized. Digital tools allow us to more closely express how our imaginations work. As Janet Murray points out in *Hamlet on the Holodeck*:

> A linear medium cannot represent the simultaneity of processing that goes on in the brain — the mixture of language and image, the intimation of diverging possibilities that we experience as free will. It cannot capture the secrets of organization by which the inanimate somehow comes to life, by which the neural passageway becomes the thought. (Murray 1997, p. 281)

The complexity of the tasks that we now find ourselves faced with means we need to respond with more complex and nuanced ideas. Our technology spaces have to provide the tools necessary to create and express those ideas. Constructing and understanding a wide range of narratives is an absolutely critical skill in the digital environment. These narratives are no longer conditioned by restrictive technologies or limited means of production. In the digital era, we must relearn how to construct our own stories in order to avoid becoming the victim of those more skilled in constructing them for us. The good news is that the technology to do so is readily at hand. The bad news is that the technology to do so is also readily at hand for others as well.

"Intertwingle" is an incredibly useful word coined by Ted Nelson to describe complex interactions between two ideas or concepts (or systems of context). It perfectly describes the complex set of relationships between narrative and technology. Digital tools mutated ancient intertwingling of *how* we communicate with *what* we communicate goes back to from their beginning. We can trace the birth of modern digital computing back to the giant code-breaking machines of World War II. These initial digital machines discerned narrative in scenarios where the participants didn't want you to understand the conversation. They made stories out of noise. One of the early explorers of this facet of the digital world was Claude Shannon, who developed the concept of "information theory" that underlies most of the modern era. Shannon was a mathematician and researcher at Bell Labs when he published, "A Mathematical Theory of Communication" in 1948 (Shannon 1948).

Shannon became interested in barriers to communication from his work in World War II, where his talents were employed making communication more intelligible and on the cryptography necessary to make those communications secure. Out of this experience, he abstracted the idea of storytelling into a system of bits, wherein the only criterion for success was the signal-to-noise ratio between sender and receiver. Shannon's groundbreaking theory digitized all communication into blocks that could humans could infinitely assemble and reassemble. His conceptualization laid the groundwork for the digital world that would evolve over the next half century. By treating narrative in this manner, he changed how we perceived the world and made possible the infinite canvases of our digital existence today. As Amber Case describes it, Shannon was "hacking reality." (Quoted in Roberts 2016) At a stroke, Shannon changed the possibilities of narrative from an analog to a digital swarm. He made it possible to hack our narratives, not just the tools that form them.

We have been exploring this "hacked" reality ever since, as it has taken over more and more of our existences. As storytellers, we are both mapmakers and explorers of this new world. Narrative conditions our reality. It bounds the possible in our minds. We do great service using technology to open new narrative pathways and imprison ourselves when we don't. "The narrative imagination has the power to play leapfrog with analytical modes of understanding." (Murray 1997, p. 282). Analog narratives force us down prescribed paths, whereas digital ones give us the capacity to create and explore new (and ancient) vistas of the mind. We now have in our hands a set of digital tools that open up to a vast narrative ✪ **SPACES** in which we can play with ideas endlessly.

These environments give us the ability to hack our stories just like we hack our tools (see Chapter 1.1).

Like everything else in this book, it's very important to boil down narrative into its most basic constituent parts in order to understand how technology changes it or doesn't. To quote Shannon and his popularizer, Warren Weaver: "The word communication will be used here in a very broad sense to include all of the procedures by which one mind may affect another. This, of course, involves not only written and oral speech, but also music, the pictorial arts, the theatre, the ballet, and in fact all human behavior." (Shannon 1963, p. 3) Storytelling is about communicating messages between two or more people. Transmitting and understanding these messages is far more important than the technology used to facilitate them, however much that technology has accelerated their frequency and range. This is not a new thing.

Industrial technologies homogenize and silo narratives. They also centralize their distribution and intensify their effect. The communication technology of the 20th century opened the door for propaganda and marketing on a vast scale. George Orwell's *1984* was merely the speculative outcome of industrialized narratives sped up to the nth degree.

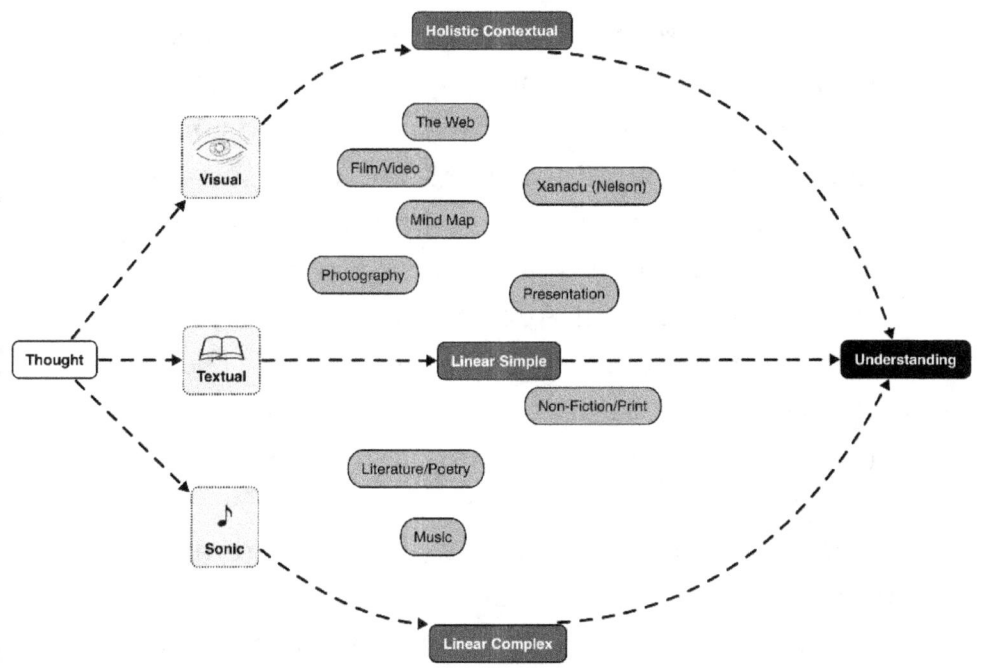
Narrative Connects Thought to Understanding

Orwell's book was ironically published in the same year as Shannon's "Mathematical Theory" that was to undermine its central premises. This is because Big Brother requires an analog flow of information to control, distort, and amplify narratives. Orwell's vision is one of a very closed information ecology. The central premises of the plot require an analog communication paradigm and that is one of its central ironies ("We have always been at war with Eurasia").

In Shannon's digital paradigm, this kind of environment could not exist because information would simply route around the wishes of Big Brother. This analog communication is not really possible in dynamic communication environments. The Eastern Bloc crumbled in the face of 1970s and 1980s communication technologies, some of it initiated by governments and some by the subnational interactions launched by the Helsinki Accords of 1975 (Morgan 2018). It would have had no defense against the digital onslaught unleashed by the internet in the 1990s. Dynamic communications environments are a hallmark of the Digital Age. As internet pioneer and activist John Gilmore observed in 1993, "The Net interprets censorship as damage and routes around it." (Elmer-Dewitt 1993) You can channel a river. You cannot channel rain.

Collectively and individually, humans are natural storytellers. This was a necessary survival trait as we passed down knowledge and wisdom to our progeny. It was our evolutionary ace in the hole over our competitors. Initially, this took the form of verbal storytelling, which was eventually complemented by graphic presentations such as cave paintings. At some point, we developed this into a truly representational set of figures that became letters and numbers and their narrative progeny, writing and mathematics. Oral and graphic storytelling didn't become obsolete, but it certainly became secondary to the magic of literacy.

Throughout most of history, the level of skill necessary to produce graphics remained fairly high. The need for advanced technical skills restricted their creation to a fairly elite group of artists, illustrators, and, eventually, photographers. When motion pictures arrived over a century ago, their technical requirements were even higher. Most humans therefore restricted themselves to writing and reading, as increasingly large numbers had access to the consumption of media through the spread of literacy. Text became the currency of information exchange and other visual expression was largely limited to the realm of art.

The aim of all communication is the development of understanding in the receiver's mind of the ideas of the sender. The ultimate canvas lies in the minds of others, not in any mediating technology. Unlike a painter's canvas, the digital canvas of the imagination flows in both ⧖ **TIME** and ✲ **SPACE**. All methods of communication seek to reach that canvas and change the perceptions of the recipient of that information.

Over the last five centuries, we have gradually trained the vast majority of humans to interact with some form of text. We made tremendous strides in literacy during the Industrial Age. However, writing is only one of many canvases available to us, and it imposes information processing biases onto any conversation. This is not necessarily bad, but it has to be recognized as intent. There are certain stories that I can tell with this book, but there are many related kinds of narrative threads that feed in and out of it. A skillful writer is a master of threading these together into a logical stream. However, even the best readers can miss a reference that seems obvious to the scribe.

The digital canvas allows us to manipulate all four dimensions of narrative (length, height, depth, and time) in unprecedented ways. Word processing and image processing are tools for idea processing.

Technology has made this easier, but we have yet to make the imaginative leaps necessary to grasp these tools of creation fully.

Our tools have always formed the speed barrier in transmitting information to others. The method that is most natural is direct face-to-face communication (assuming a common language tool), but this is inefficient both temporally (it is fleeting) and from a distribution perspective (limited by the reach of a voice and the attention span of the listener). Writing thoughts solves both problems but imposes its own limitations. Text creates a linear narrative with a beginning, middle, and end. Effectively communicating this way requires years of practice. It also creates a straitjacket of thought that moves in a singular direction.

Writing is highly useful in the linear kinds of problem solving that have characterized human activity since the beginning of the Enlightenment. However, it limits our ability to analyze the trajectories of the non-linear problems that are increasingly the norm in today's world, such as global climate change, social injustice, and even "fake" news itself. Today's stories lack a beginning, middle, and end. Industrial, analog information ecologies simply can't cope with the complexities of these kinds of problems (more on this in the next chapter).

The Digital Age gives us access to a whole new range of storytelling tools, from the intertwingled chunks of information that Ted Nelson envisioned in his Xanadu project to the importance of coding expressed by Kay, Papert, and others. (Nelson 1974, Kay 2010, Papert 1980) These modalities have one thing in common. They perceive information as Shannon did; not as an analog stream, but as a constellation of bits of data that need assembly to turn them into coherent ideas. In this reality, our stories become multidimensional worlds. Perceiving this reality effectively can have a profound impact on how we problem-solve and process information. It can literally change how we think. In a 2007 study, Yoram Eshet-Alkalai and Nitza Geri looked at the intersection between critical thinking processes and digital versus analog information consumption. They found that,

> Unlike the linear reading and learning dictated by a printed text, the digital, hypermedia-based text requires the reader to master a high level of non-linear branching and thinking capabilities Smilowitz, Lazar et al., Lee and Hsu, and Rigmor and Rosemary describe the abilities to create mental models and metaphors as crucial for "surviving" the intricate hypermedia structure of information representation on the web, and employing critical thinking skills. In most of these studies, the researchers found that younger people are more effective non-linear Internet learners and readers than adults. (Eshet-Alkalai and Geri 2007, p. 276)

What this seems to show is that the linear mode of thinking, and, by extension, linear problem-solving, is a learned behavior. Like every tool, the linear approach has its utility. It produced 300 years of the most intense scientific and technological growth in human history. It has also led us down a set of paths that have had unforeseen consequences, including overpopulation, pollution, extremism, and war. We need a digital approach to world building in order to analyze problems at all levels holistically; to hold tertiary effects visible; and to create solutions based on unexpected intersections of information and expertise. Digital storytelling holds the key to our political future.

All storytelling is world-building. In its purest form, this results in poetry, followed by fiction. However, even this nonfiction book is creating a world of ideas and thoughts that are being laid out by my prose. Over the course of it, I have struggled against the edges of this world because it depends on the linear modalities of text. It is difficult to avoid becoming an informational tyrant; to dictate the terms of my arguments instead of letting them become a base for a creative stew of ideas. I have attempted to break out of this trap in part by adding graphics and idea mapping at the end of each chapter. Graphics allow me to show the same concepts in ways that textual tools do not allow me to do and prompt a different level of reflection by the viewer. Creating these graphics would have been technically difficult without the digital tools that I had at my disposal. The tools at hand constrained and conditioned the informational and conceptual world that I imagined, but my tools were themselves also limited by my mental capacity to process and employ them effectively.

I could actually take this much further in the Digital Age by completely reshaping the digital world that you, the reader, are traveling in. This is precisely what Kay, Papert, and Nelson have been imagining for decades. They argue that coding is a crucial part of the digital reinvention of possibilities. (Kay 1992, Papert 1980, Nelson 1974) Coding literally allows you to create worlds that move far beyond the conceptual imagination and to build tools that propel you through entirely new landscapes of the imagination.

However, despite the best efforts of Kay and his team at Viewpoints Research Institute, even today coding at a complex level remains a challenging task from a purely technical perspective. The word processing program I am writing on right now, to cite just one example, required millions of lines of code. (see Kay et al. 2010) The conceptual design of this book (and indeed all books) is to provide the reader with a sense of the possible and shift his or her attention away from the technicalities involved in expressing and analyzing ideas. This frees the reader to engage in the much more difficult parsing of the ideas themselves, while employing new tools to create new perspectives.

We can take this all the way back to the basic level of text. Even on the textual level, Digital Age tools can change the parameters of the building of worlds profoundly. As early as the late Seventies, about the same time as I first got involved in computing, science fiction authors, in particular, envisioned computers as "world-building" devices. According to Kirschenbaum, both Jerry Pournelle and Frank Herbert grasped early on the potential of the computer for "world building." (Kirschenbaum 2016, pp. 101-102) While both of them looked at this potential as applying specifically to designing the worlds necessary for their work, conceptually world-building is world-building. It doesn't matter if your goal is creating a coherent alien landscape or imagining an alternate reality where we can attack the thorny problems we encounter daily.

World-building also changes how we think about narrative and shifts it from a linear model of storytelling to a design process for ideas. Alex McDowell, who was the production designer for the film *Minority Report*, developed his world building "mandala" as a mechanism for building a science fiction world around Phillip K. Dick's short story of the same name. The timing of the schedule forced McDowell and script writer Scott Frank into a process where the script didn't dictate the movie. Instead, they could develop the production design and have the script grow organically out of that process. McDowell developed his "mandala" out of this process.

Prior to *Minority Report*, most production design focused on the "hero's journey." The main protagonists moved through the film and the production design team constructed the world, as necessary, around their movements and the overall plot of the story. What was different about McDowell and Frank's process was that McDowell convened a team of futurists and they mapped out the world before they knew the characters' actions in the world. Only then, Frank wrote the more traditional narrative within that "story world" with all the pieces of the world already in place. (Von Stackelberg and McDowell 2015, pp. 25-46) Much more in keeping with real world narratives, the story filled the spaces between the concepts, not the other way around.

One of the major struggles we have in facing complex problems is one of perspective. It's hard to remove ourselves from our own "hero's journeys" and to see the world from different viewpoints. Linear text has a tendency to lead the hero's journey style of narrative. Problems are simplified or discarded in the service of creating a coherent story. (I'm doing this now.) The problem is that the world is rarely that linear and we almost always introduce biases toward existing norms. Visually separating yourself from what you assume to be a logical path (and therefore true) forces you to look beyond built-in assumptions about the world in front of you and consider the world to the sides of your path.

It is essential that we develop a facility for stepping back and looking at the frames around our "logical" constructs. These frames may come from mental blinders or they may simply be a product of a lack of relevant information. Digital tools can help with this task by making visual expression more accessible. In his deconstruction of comics, Scott McCloud makes a powerful point when he observes that the absence of information between two comic panels makes the creator makes the reader an active participant in the narrative. The reader's imagination takes over and makes a myriad of connections. (McCloud 1993, pp. 66-73) He even goes so far as to argue that the images don't even have to be related to one another in any obvious way.

> No matter how dissimilar one image may be to another, there is a kind of alchemy at work in the space between panels which can help us find meaning or resonance in even the most jarring of combinations.... By creating a sequence with two or more images, we are endowing them with a single overriding identity and forcing the viewer to consider them as a whole. (McCloud 1993, p. 73)

Concept mapping allows us to shake up these perceptions in the manner that McCloud describes. By plotting "panels" on a page, our brain makes connections between them instinctively, even if they are seemingly unrelated. I often combine these kinds of exercises with lateral thinking cards to make the exercise even more challenging for those engaging in the activity. Reverse engineering the McDowell Mandala and combining it with concept/mind mapping software has allowed me to develop a way to help my students and workshop participants to deconstruct problems. The Digital Age technology tool of concept mapping, combined with insights about how the human mind and imagination work, has facilitated my ability to get humans to think differently about their world and circumstances. McCloud provided me with the critical insight, but McDowell provides me with a set of rules that gives a useful structure to the canvas.

The core of McDowell's system comprises three axes. First, we have to ask how we construct mental maps around a challenge. What biases and preconceptions do participants in our world bring to the table? For instance, take the assumption that crime is prevalent in our society and that we live in a dangerous place. Media portrayals of the news reinforce these perceptions, but they form a fundament for understanding how we perceive the problems of our society and how we prioritize them over others. Second, we must then ask how we organize ourselves around those issues. Do we develop systems of prisons and law enforcement, or do we develop systems of hospitals and job training? Do we prioritize these things over other things such as healthcare? This is the Social Map Axis of the map. Finally, there is what I call the "you can't fool Mother Nature" axis of the map. This is the Resource Map and refers to everything from the limitations of funds for building prisons, the cost of housing prisoners, and so forth. In other contexts, it can refer to scientific realities, such as climate change or demographics. This is my adaptation of McDowell's map using a concept mapping software program.

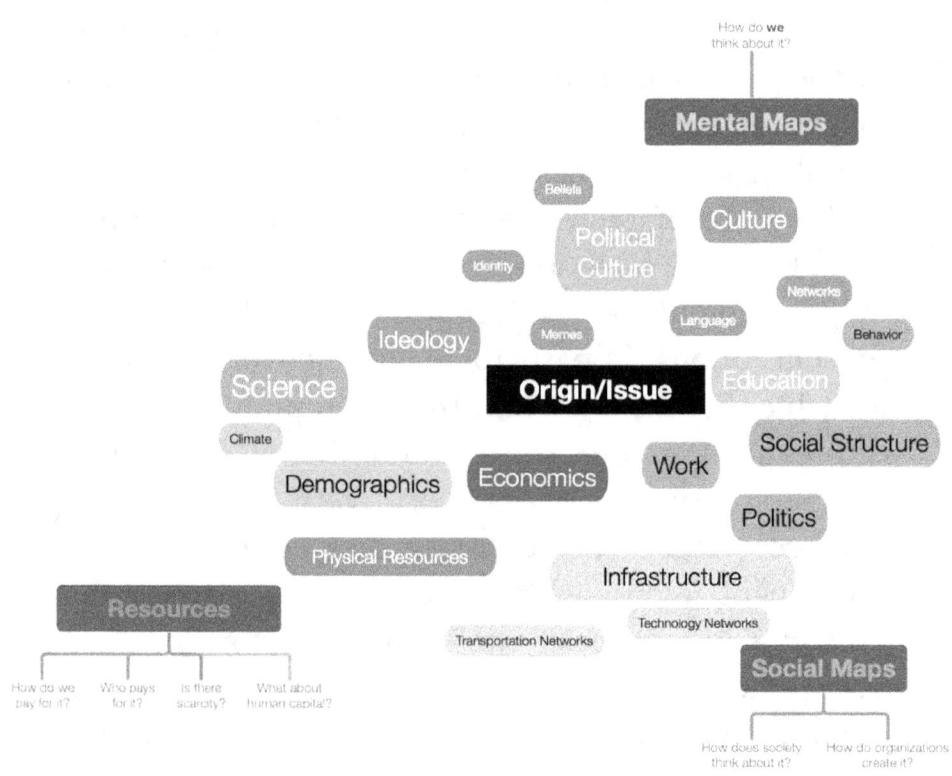

McDowell Mandala Used to Visualize Politics

Besides giving us new perspectives for deconstructing problems, this is a dynamic map and we can manipulate it live. The map stimulates conversation. During class, I ask students to bring in a selection of random sources as their homework assignment. We then take those sources and place them in various places on the grid. Some sources may fall into multiple areas. This allows us to explore relationships between how we perceive issues, how we organize ourselves around those issues (in this case using governments), and the realities of resource allocation around those issues. I can challenge them when they bring in sources that they think are resource questions, but then turn out actually to be *perceptions* of resource issues. In this way, I help them work their way through their design challenges as a set of exercises that add reality and meaning to the otherwise abstract concepts you have to deal with in a course on political science.

This same technique can be used to brainstorm ideas in workshops. The map is adaptable depending on the subjects being discussed. By using this technique, facilitated by technology, I can facilitate the reshaping of narratives dynamically in a live session. Technology allows my students and participants to deconstruct their realities in hitherto impossible ways. In my class, we use a program called **Miro** to do this exercise. **Miro** is an online application, which my students can access live in class but is also persistent. They can continue to iterate and reference it as we move through the class. Miro creates a living sandbox. Participants can access the map anytime, and it can form the backbone of the narrative we collectively construct between live sessions. By giving us a dynamic canvas, concept mapping has changed the IdeaSpace of our conversations in much the same way that word processing changed the IdeaSpace of my writing.

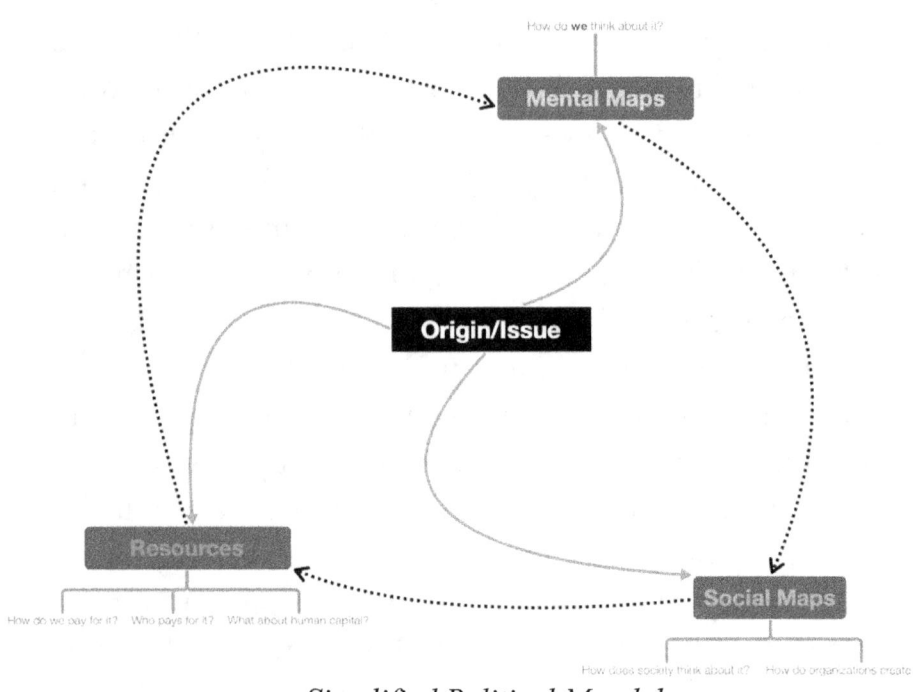

Simplified Political Mandala

I use concept mapping extensively as a tool to help people visualize problems and develop strategies for engaging with them. McDowell's Mandala is just one strategy. I have developed complex systems thinking based on Donella Meadows's *Leverage Points* concept into a brainstorming game using the same software. (Meadows, 1999) This allows me to work groups through all the systemic barriers that may face executing a particular project or initiative.

While Meadows created her *Leverage Points* in an analog form, it lends itself well to visual exploration and adaptation as a tool for problem-solving. (I explore this in greater detail in my gamification of Meadows's system in Chapter 2.3.) McDowell also started in a somewhat analog form, but he quickly developed his mandala into a graphic form to teach other production designers this technique.

The accessibility of this process is a fundamental product of the digital age. A decade ago, I would not have been able to create an interactive concept map for my students. Even with my advanced technical skills, I would have found the creation of my *Leverage Point* game board would have been considerably more difficult in the 2000s. The dynamic nature of the concept mapping software allows me to explore new configurations of complex ideas and to make them more accessible to lay audiences who could benefit from their power but do not have the time to explore them in more traditional narrative structures. This is the power of digital narrative spaces.

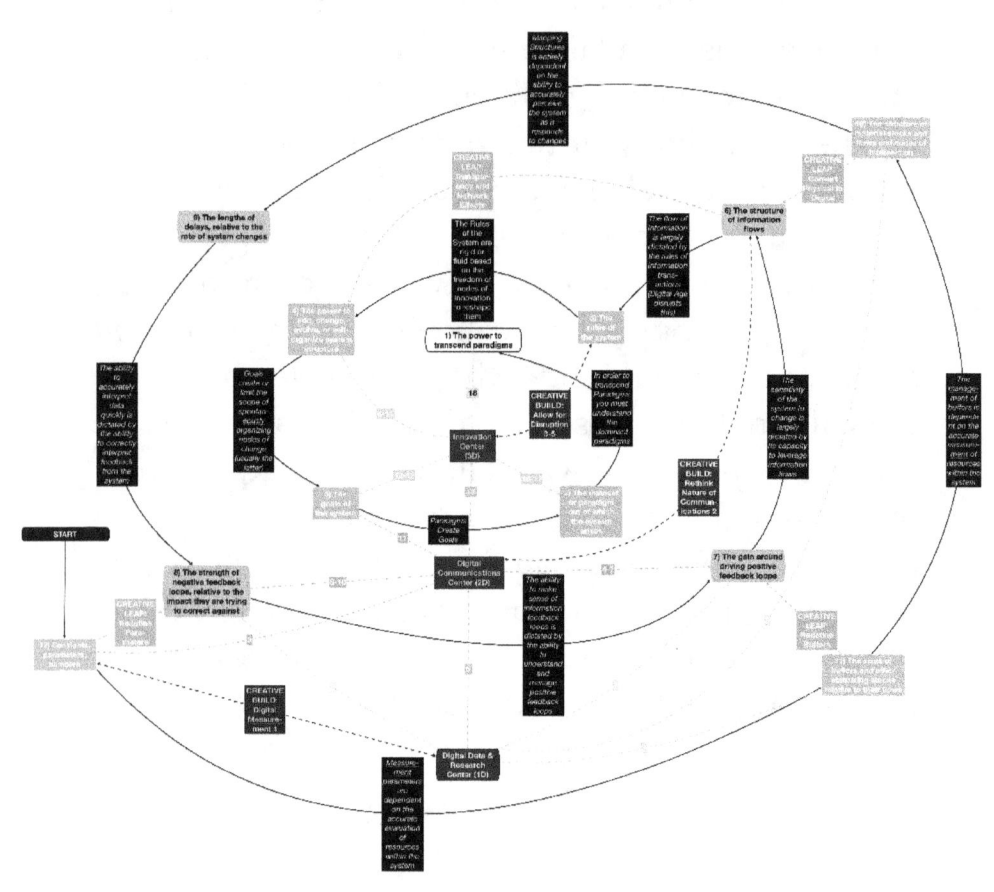

Game Board Constructed from Meadows's Leverage Points

The computer allows us to experiment with ideas in a way that older technology did not. The artistically inclined could do this with a sketchpad, and there was always the ubiquitous blackboard, but these were imperfect tools for truly reshaping ideas. World building goes back much further than McDowell, however. In a non-graphical and eventually semi-graphical expression of world building, one of the early "killer apps" for the personal computers of the 1980s was VisiCalc, a forerunner of Excel. It was, simply put, a spreadsheet application and was incredibly useful as a bookkeeping instrument. However, its users quickly discovered that it had a hidden power: it could model out the future. Users could map out scenarios with numbers in the sheet quickly and easily experiment with outcomes.

Numbers are abstractions of ideas in the same way that letters are. Linked numbers make formulas that form their own stories in the same way that linked letters make words that make sentences, and so on. What computers did in software was to disassociate the ideas they represent from the physical processes necessary to produce them. This allowed hitherto unimaginable levels of manipulation of these ideas. The problem was wrapping our heads around the metaphors necessary to express them. We still think in terms of spread*sheets* and *pages* when those metaphors really don't work in the same way anymore.

One of the critical mental leaps we must make is to think of information as networks of ideas instead of linear arguments. Looking at challenges this way, we can focus more attention on how we connect those bits of information to one another. This is as important a part of the creative process as the physical creation of media. Narrative exploration easily leads to play as a way of scenario-building and exploring structural change. We will discuss this impact of the Digital Age more thoroughly in Chapter 2.3.

The concept of networks of ideas is something that stimulated the very computing revolution that we are dealing with now. Vannevar Bush's 1945 article, "As We May Think," was a manifesto for the Information Age long before we even knew what that term meant. Bush, who headed President Roosevelt's science projects during World War II, wrote in *The Atlantic* that the challenges facing vast projects such as the (unnamed and still secret) Manhattan Project, required a novel approach to managing information. (Bush 1945) Bush's speculative proposal, *The Memex*, inspired the developers of the internet from Doug Engelbart to Ted Nelson to Tim Berners-Lee to apply Shannon's digital vision to Bush's prescient understanding of the limitations of analog storytelling.

The critical element that was lost in Berners-Lee's execution of the web was Bush's concept of "breadcrumbs" that allowed the user of *The Memex* to backtrack through blind alleys and explore abandoned or redirected sources of information that might have become more significant in retrospect. In creating the network, Berners-Lee's system inadvertently buried the creative connections as an ephemeral technical abstraction invisible to most users. The technical constraints of 1989 imposed these kinds of limitations on his innovation. Unfortunately, much to Berners-Lee's frustration, this opaque system of connecting information has persisted to this day.

In the 1960s, Ted Nelson began trying to build these kinds of creative connections but has struggled to realize his (and Bush's) aspirations because of the two-dimensional nature of our display systems. He has been trying to realize "breadcrumbs" ever since. A variety of barriers, ranging from funding to indifference, have stymied Nelson's Xanadu project. (Nelson 1999) At its root, however, Xanadu, while solvable in 2D tech ✪ **SPACE,** is conceptually difficult to follow. The fundamental problem remains one of creating a visible conceptual network out of a sea of information networked together with no conceptual framework. Perhaps as we reimagine informational connections in a virtual reality environment, Nelson's concept will become more intuitive.

As a result, technology has enabled an explosion of disconnected narratives lacking the contextual power of breadcrumbs. Breadcrumbs are the paths of digital mapmaking. It is easy to get lost without them in a sea of information. Without them, information becomes randomized, lacking in context, and subject to abuse. With them, there is tremendous potential to understand the world in completely new ways. We will look at how revisiting breadcrumbs creates opportunities in the case study of "Knowledge Networks" in Chapter 3.3.

Centralized, analog streams of narrative created artificial norms of understanding that persist in the digital world. The industrial press, the radio, and television followed text. These media have two things in common. They established dominant narratives, and they confined those narratives to a relatively small elite. Now it is at least possible that we can judge the *quality* of ideas rather than their access to the corridors of power. The internet undermined dominant narratives because it democratized the means of distribution. You no longer have to have a printing press or transmission tower to get messages out to the masses. Simultaneously, the means to create non-textual forms of media have exploded.

It is easy to lose sight of the basics of communication and to feel disempowered in our ability to communicate with each other. The democratization of content creation has brought with it unprecedented noise, which we will explore in the next chapter. We have created tools that have lowered the barriers to entry to those wishing to communicate in ways that do not conform to traditional textual means. As creators and transmitters of information, this should be profoundly empowering. However, even if the tools are there for us to create new pathways of communication, it is still up to us to grasp them and unlock their potentialities.

We should not underestimate the task before us. It will be necessary to master novel forms of digital tools (visual or otherwise) to navigate in the informational world we have entered. The complexity of the information environment that spurred Bush to write "As We May Think" has continued to evolve far faster than our imaginations. We need imagination tools to unlock new ways to create and consume narrative. Adaptation, both as creators/hackers and navigators/curators, will take ⧗ **TIME** and massive intellectual effort, but we increasingly have the tools to engage in that project.

This project will require an effort of mapmaking akin to that facing those exploring the world as we were just developing the technological tools necessary to map it accurately. Mapmakers of the 17th and 18th centuries (and beyond) undertook tedious, often lifelong, projects to make precise measurements of our terrestrial landscape using primitive tools of measurement. Often, they were literally venturing into conceptual *tabula rasas*. (Wilford 1981)

The struggles of ancient mapmakers have a direct connection to the digital narratives that have opened before us. In much the same way as they sought to represent stories of the physical world through maps, we must now create maps of the conceptual narratives and environments opening up to us. Anyone interested in understanding the current struggles we have with creating coherent narratives should also consider the challenges we have faced in our past in creating accurate maps and visual representations of understanding.

It's not something you think about in our context, but the digital world we are contemplating resembles the world as it appeared to a 16th century cartographer. It's useful to make the analogy and imagine for a moment making a map without the aid of modern technology. In an age of Google Maps, laser spectrometry and range finding, it's hard to make the mental leap to a navigator on a swaying ship trying to time the passage of the moons of Jupiter (which is almost impossible) in order to figure out his longitude, much less draw an accurate representation of where he'd been. Even the act of mapmaking on the ground was a tedious project involving lifetimes of work just to map a single European country. Four generations of Cassinis worked on accurately mapping just France over the course of more than a hundred years (1670s-1780s).

Ancient mapmakers labeled areas they could not map "terra incognita" or incomprehensible earth. They also often embellished these areas with fantastical beasts, including the occasional dragon. It is sobering to realize that it was only recently, within my lifetime no less, that we have finally slain all the "real" dragons out there. For much of human history, the world just petered out into the unknown. As recently as the 1970s, the United Nations, which acts as a clearinghouse for maps based on the scale of 1:1,000,000, noted that we still hadn't mapped half of the land area of the planet to that scale. It is even more sobering to realize that the science of mapmaking is considerably faster than cultural cognition of mapmaking. While we may be relatively quick to accept Google's instructions on how to get to the nearest Starbucks, the cultural and geographic connections we have with Mexico or Canada elude many Americans.

What does this have to do with narrative? We are digital mapmakers on the shore of a vast sea. Some of us are venturing out with compass and sextant, but we still ground our cultural contexts in analog soil. Where are the hidden shoals that need to be mapped? Are there dragons out there? Just like the fictional dragons on these maps, the digital dragons we discover will live in our imaginations. Even more than in the physical environment, most of us don't even realize that there is another reality to be explored or may only perceive it vaguely (and with dragons).

Like the early mapmakers, this task will involve painstaking work and the development of whole new sciences of digital location and identity. We are going to need novel sets of tools to navigate in this territory and to make maps for others to find us. We have to develop our own tools in order to measure the geography of these places that we have so boldly ventured into, if for no other reason than to show the rest of the world that what we have found is safe and can be immensely beneficial to humanity.

Just as latitude and longitude overlaid geometry onto our real world, we must find abstract analogies in relationships devoid of spatial form to understand where we stand and where we can go in the digital world. Shannon may have shown us the way with the mathematics of communication, but we will also need social science to overlay human latitudes and longitudes onto the new cultures, communities, and relationships that will make exploration of the digital landscape possible. Maps are a blank canvas waiting to be filled. Exploring and mapping the digital canvas of ideas will be as big of a challenge as those facing the explorers of our physical world, if not more.

Mapping the digital canvas will require the redefinition and development of entirely new literacies that allow us to explore these new worlds in the same way that mastery of the sextant, the theodite, and chronometer allowed mapmakers and navigators to map the world in the 17th and 18th centuries. The computer gives us our 21st century chronometer to shift time and perspective. "Not only is the computer the most capacious medium ever invented, but it also allows us to move around the narrative world, shifting from one perspective to another at our own initiative." (Murray 1997, p. 283) It is only with these tools that we will achieve the perspective and new cultural contexts necessary to navigate our way out of the information morass we currently find ourselves in.

A canvas has suddenly opened before us over the last 30 years of digital transformation. The barriers to entry for graphic and cinematic storytelling have rapidly dropped through the elimination of cost and a sudden ubiquity of tools as simple as the cell phone. This sudden proliferation of digital tools has suddenly given many people access to a wide range of storytelling media with no real understanding of how, to paraphrase McLuhan, the medium reshapes the message. Chief among these is the sudden accessibility of visual communications. Writing is an inherently linear process. Words, sentences, chapters, and books have an inevitable beginning and end. This is a product of the visual representation of that information through letters, words, and phrases. Still images are bounded by frames. Movies, as we understand them today, combine that framing with the linear path of text. The web, however, is a constant living, breathing entity of stories.

A firm grasp of visual and all forms of narrative literacy is critical to map how we communicate today, intentionally or unintentionally. (Eshet-Alkalai and Geri 2007) As Marshall McLuhan said in *The Medium is the Message*, "The serious artist is the only person able to encounter technology with impunity, just because he is an expert aware of the changes in sense perception." (McLuhan 1964, 2003, pp. 207-208) My career as a photographer has unquestionably changed my approaches to problem-solving because it has taught me to perceive my working canvases differently than someone who only communicates in text. It is essential that we educate people in visual thinking and communication in today's world, both as consumers and creators of ideas.

Technology inevitably forces us to reframe our perspectives in ways that are uncomfortable to those who are used to the linear forms imposed on us by industrial and textual narratives. McLuhan dissected this mode of thinking in much of his work, which was written at the beginning of the electronic age and described how the emerging media of the time, especially visual media, had reshaped the linear textual landscape.

As McLuhan points out, the Industrial academic and literary elite saw the book as the highest form of art and have continued to perpetuate that elitism. (McLuhan 1962, 2003) I do not mean to dismiss or disparage the power of the book (as I am writing one literally right this moment) but its primacy as a means of communication is a product of the technical limitations of analog communications environments.

Over the centuries, we elevated the book to be a bastion of the establishment. The elite used the power to control discourse instead of liberating it. Mass production of books during the Industrial Age created an environment that spread reading to a vast number of new readers while acting to constrain disruptive ideas that were not perceived as worthy of widespread publication. This side effect ran counter to the effects printing had during the Age of Enlightenment, where printing was used to elevate ideas, not individuals. The Industrial Age had the effect of promoting a hierarchy of ideas and the process of publication was integral to establishing an elite whose ideas we perceived as being superior to others. If you were worthy of publication, then your ideas were superior as well. As Tom Standage has pointed out, "those in authority always squawk, it seems, when access to publishing is broadened." (Standage 2013, p. 244) Now, besides democratizing writing, we have literally exploded the canvas of storytelling by giving people tools formerly reserved to technical specialists. Then, we compounded the disruption by giving the masses the means to distribute the products of those efforts.

Visual communication threatens textual monopolies of thought and so the traditional elite looked down on that medium. The internet and tools such as YouTube have undermined this school of thought, but it is deeply ingrained in our social structures. We don't teach our students how to visualize because the vast majority of teachers went through textual training in their undergraduate and graduate experience and don't even know how to approach visual literacies, much less create in them. My photography aside, that was certainly the case for me.

You see this deficit in a vast range of areas, from incomprehensible slide presentations to textbooks littered with confusing visualizations. There are exceptions to this, and some disciplines are more forced, by necessity, to visualize their content. Outside of dedicated programs, however, there is little training in visual information creation and presentation. "Technology" programs often teach the mechanics of using software such as PowerPoint, Illustrator, or Photoshop, but there is often little thought given to the purpose and the power of these software packages to reshape narrative.

As I illustrated with my adaptation of McDowell's world building mandala, graphics created through concept mapping software are a simple technologically enabled way to create a diagram or idea flowchart that can augment our stories, either in presentations or through various online platforms. We can even create concept maps live during brainstorming sessions and meetings to capture the complex interrelationships of the discussions that happen and that we often lose in the fragmented, linear nature of meeting minutes and notes. They present a way of deconstructing and reconstructing problems that are often lost through analog text interactions, such as email chains (the modern equivalent of dueling memos).

Adding graphics would seem to be a simple solution, but, employed randomly, they do not increase the canvas of ideas we are trying to communicate. We often add visuals indiscriminately without thinking visually. We have all sat through bad slide deck presentations where the presenter completely misses the point of having slides. The slides are there to form a backdrop for the presenter. Used effectively, they can complement the other performance aspects of the conversational ✿ **SPACE** they take place in. More often than not, they are an ineffective distraction from what information the talk may contain. Taking a class or training on the software rarely teaches us this level of complexity. Most classes that "teach" presentation tools, such as PowerPoint, gloss over the narrative changes that it imposes on us through its transition from a linear textual narrative to a linear visual narrative. They also frequently miss the information transfer capacities of alternative visual media and how to incorporate them into the flow of your presentation narrative.

Edward Tufte pointed out almost twenty years ago in an influential Wired article, "the PowerPoint style routinely disrupts, dominates, and trivializes content." (Tufte 2003) This is a consequence of a poor understanding of the narrative canvas and PowerPoint's place in it. Slide decks are a performance augmentation tool. A good analogy is that it is the set on which the play is staged. We would not expect the set to carry the story and, in isolation, it's a pretty poor substitute for the performances that take place there. Therefore, at a minimum, slide presentations need to be augmented by more persistent visual and textual media. I've worked around this by creating websites for my presentations as a mechanism for glossing my presentation to provide contextual linkages. I consciously designed this strategy to create a persistent, living complement to what happens live.

Part of visual literacy is understanding how various visual media (slide decks, mind maps, graphic designs, etc.) fit onto the larger media canvas and deploy them mindfully. They are powerful tools when applied correctly, dangerous tools when used to manipulate, and comic tools when applied incompetently (which is so often the case these days). As recipients of information in Shannon's model, we have to be mindful of how we can use it for enlightenment as well as propaganda. As creators, we will be more successful in building the world of our choosing if we can understand and use the tools available to us to change our perceptions and craft solutions to the massive challenges facing the world today.

The Digital Age opens up a panacea of communication tools for us. This can be both liberating and intimidating. It is easy to get overwhelmed by the complexities of visual communication and visual artists spend years honing their crafts. The need for these kinds of professionals will not go away and there is a real shortage of talent in this area, which is often viewed as a "soft" occupation.

Graphic experts, like my friend Karina Branson, will be even further liberated by the tools at their disposal. One skilled, imaginative human with a marker and a roll of butcher block paper may be all the technology you need, but that same talent may reach new heights with an iPad and screen sharing software. However, technology allows us to begin these explorations without such an expert in the room. I have certain talents, honed by decades of hard-won experience, in the visual area. Drawing is not one of them. However, with concept-mapping software, I can express ideas, provide platforms, and stimulate conversations that were impossible for me using older forms of technology. I can also explore concepts visually and rearrange ideas as easily as I write these words on my computer. This journey started with the textual rearrangement made possible by my Apple][+ and Word Handler all of those years ago.

Technology challenges us all to be more reflective as we tell stories. I am not terribly enamored with the term "digital literacy," which we will explore more extensively in the next chapter. "Literacy" is an all-encompassing term that now needs to be understood in the broadest possible terms. We have to learn lessons that were previously only necessary for serious artists and intellectuals. McLuhan understood that fundamental fact as he explored "the medium as the message" in the 1950s and 60s. How we communicate our stories matters. Some stories are better told visually. Some stories are better told textually. Some stories are better told musically. All hold equal validity. They are just sets of tools. What the Digital Age has done is to change the accessibility and reach of these narrative spaces, not their fundamental realities.

As the world gets more complex, visualization becomes even more critical to our methods of teaching, learning, and all other forms of storytelling. The trick is not being afraid of it. Get out the finger paints. Try to tell your next story through pictures. If you mess up, play with it and refine it just like you would with text. Above all, apply a visual eye to your presentations, websites, and videos. Be aware of the narrative shifts created by others' use and misuse of the same tools. We are all storytellers. That hasn't changed.

In the next chapter, we will explore another meta media, the Network. Applied effectively, it complements the canvases we create individually. The

Network is nothing more than another narrative canvas. Instead of being our individual tool for communicating with each other, it represents a collective platform for storytelling. Again, this is not a new thing, but Digital Age technology enhances and expands it in much the same way as it has expanded non-textual communications. Like our individual efforts, it is a product of digital finger painting on a vast scale. However, the mere fact of that scale raises the stakes of that activity from ☼ **SPACE** to ⌛ **TIME**. Everything digital is copyable and distributable, instantaneously. Unfortunately, humans don't operate instantaneously when processing ideas. Fortunately, the Digital Age also provides us with the tools to filter and process those ideas into a human timeframe.

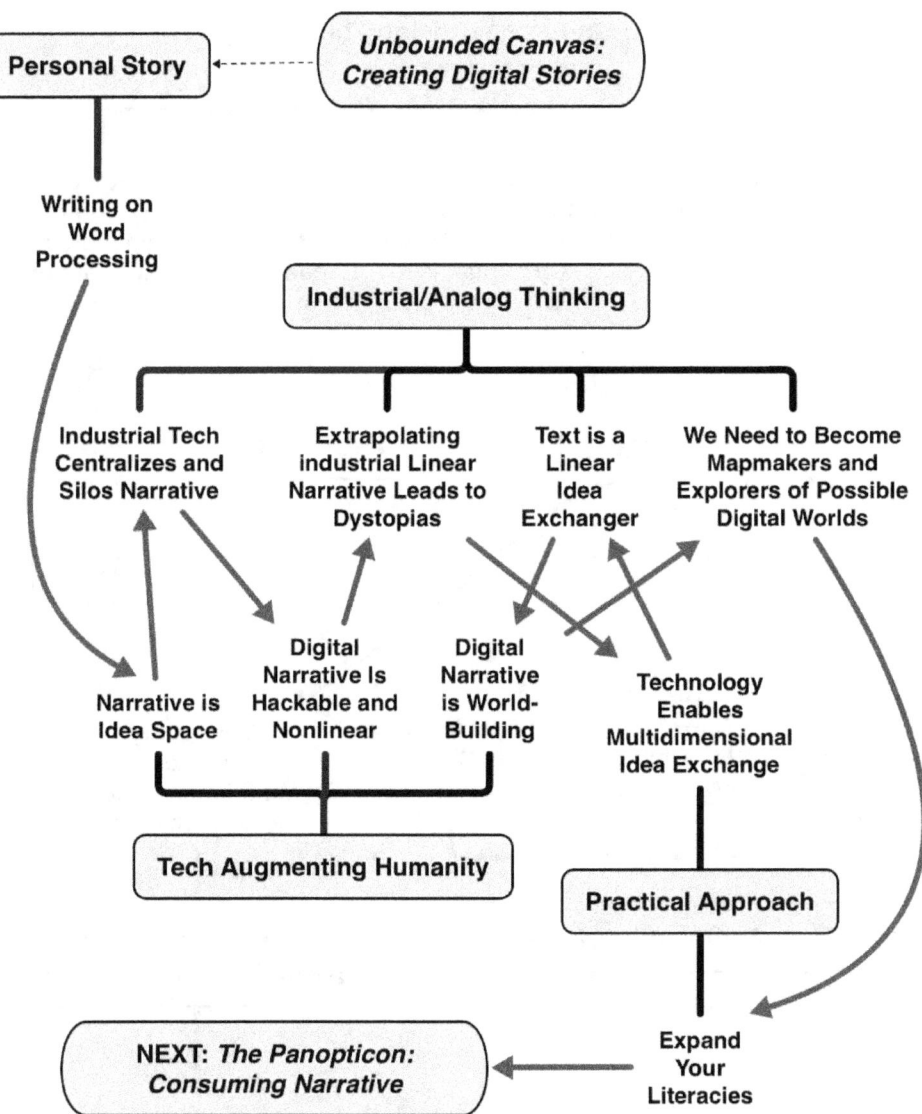

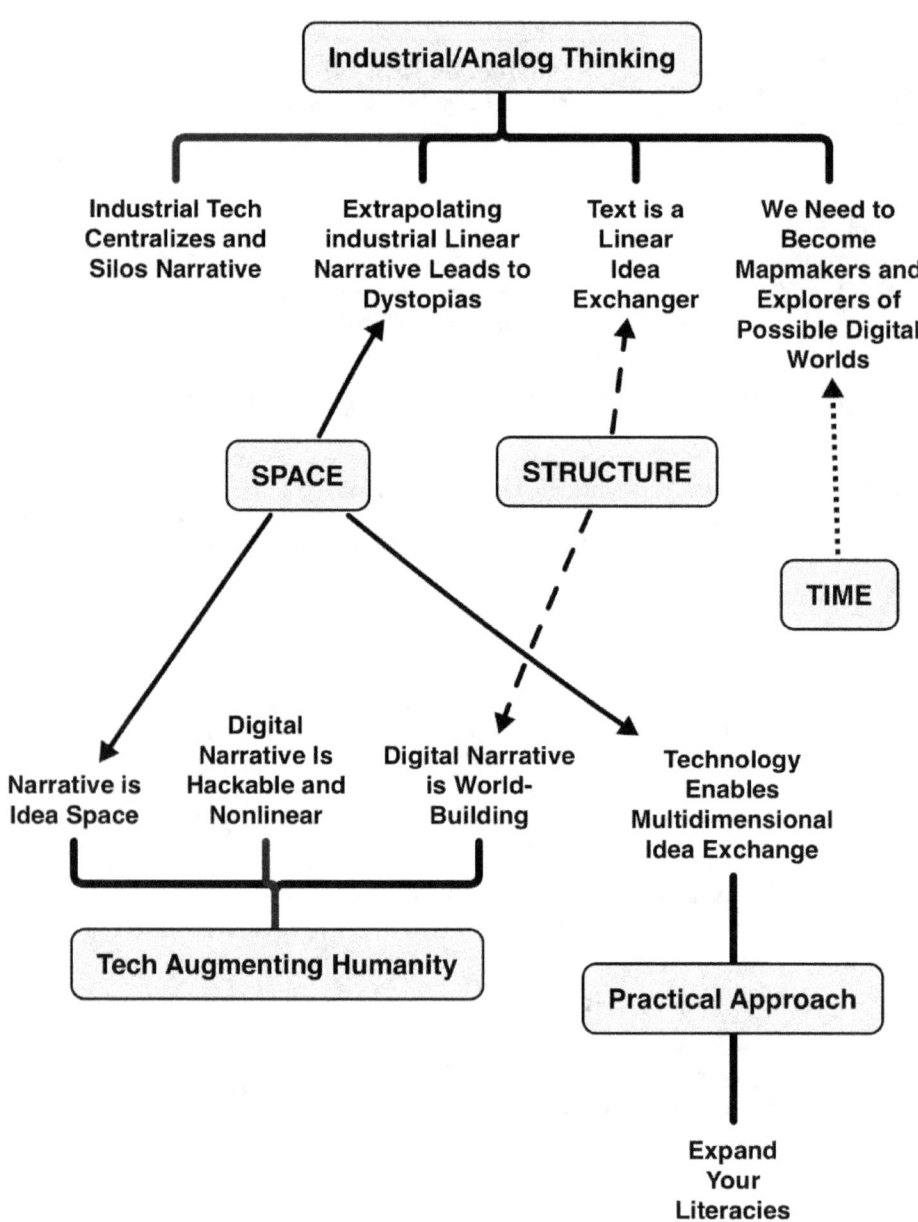

⌛ TIME

The Network that humans have created out of the Internet and its related technologies has bent ⌛TIME in ways that we are still coming to grips with. The struggle here is with ⌛TIME itself and carving out reflection ⌛TIME in a world of informational overload. Information works and flows in the Digital Age differently. The key elements in coming to grips with managing our ⌛TIME in the Digital Age are our capacity to create transparency, curation, and perspective. Our ⌛TIME is ours. We are a crowd of individuals, not individuals in a crowd.

2.2 Living in the Panopticon: Consuming Narrative

You can always get what you want, but you can't always get what you need.

Gary Kamiya, 2009

You can't always get what you want; but if you try sometimes, you might find; you get what you need.

Jagger and Richards, 1969

I was a graduate student in the 1990s when I first encountered the implications of the network firsthand. It was at that point that I discovered a group of Listserv groups operated out of the University of Colorado. As someone who was exploring the borders of my discipline (political science), the idea of participating in geographically distributed discussions seemed like an alluring proposition. One attraction of these groups was that they were extremely leveling. No one knew whether you were a tenured professor or a lowly graduate student. The concept of being evaluated on nothing more than the quality of my ideas attracted me. It immediately struck me as being a very useful community-building tool that would expose me to a range of new thinking and make my own ideas stronger.

The reality quickly became something different. Instead of becoming a venue for the exploration of ideas, the fora quickly became arenas of policy disagreements, personal attacks, and, sometimes, outright disinformation. Being able to float and dissect controversial ideas is a cornerstone of academic discourse.

Unfortunately, these online environments became early bastions of normative thought and speech. Instead of a graduate seminar, which is what I was expecting, the relative anonymity of the participants made the conversations more akin to Lord of the Flies. Hot takes, flaming, and other reactive forms of communication often came to dominate the discussions because of the speed and scale of exchanges made possible by the technology. This was not because the participants were necessarily bad people, but because the speed at which the narrative could suddenly travel inflamed minor human failings into major points of conversation. In the intervening 30 years, we have scaled this experience so that now almost anyone can experience its joys.

I, however, learned from the experience and I've avoided the large-scale wars that seem to occur daily on Twitter and Facebook by being very sensitive to the amplification effect and ⏳ **TIME** distortions created by the speed and scope of our new communication tools. This amplification of culture is a product of the speed and distance with which we can communicate using the network.

The new realities of digital ⏳ **TIME** have shifted our collective communication. It is up to us to shift our conceptual approach to these tools and use them for the possibilities that they hold instead of letting them stimulate the worst in us. We can use a hammer to build, or we can use it to kill. The same is as true of this set of technologies as any other we have discussed in this volume. We must shift from the reactive to the reflective to create civilized discussion instead of acrimony. Speed kills in information flows as much as it does on the road. Our rediscovery of digital humanity is contingent on our ability to maintain perspective in the face of the ⏳ **TIME** and ✿ **SPACE** distortions that the network has created.

Perspective means that we need to adapt our vision to the tools we have created. We need to step back for a moment and consider what the network actually is. It is so much more than the sets of routers and switches that deliver content into our lives. It is also not new, although the media likes to portray it as being so. We have built the network upon the foundations of our existing, usually linear networks. This creates distortions. Human communication is a complex thing. We have scaled it to an unprecedented degree in the brief timescale of a little over two decades. The result has been the sudden release of a cacophony of hitherto unheard voices, some long overdue, some repressed, and some we thought buried. It is no longer possible to impose a normative understanding on these flows of information and any attempt to do so risks oppression and censorship of ideas.

This is not an argument for relativism, but an argument that linear thinking, a product of industrial, textual narrative structures, doesn't work well in this environment. Instead, we need to develop new ways of looking at human narrative output. A line is the wrong metaphor for narrative today. Digital narrative is more likely to form a scatterplot. In a scatterplot, the challenge is to make our own patterns and look for ways in which the plot itself forms a useful narrative.

We have had networks since the first humans sat around a campfire and exchanged ideas about how to hunt more effectively. These networks serve as "knowledge amplifiers" as humans have shared experience and knowledge throughout the ages. We facilitated that sharing through the invention of literacy and then the printing press, but "word of mouth" has remained a critical avenue of sharing. What the internet has changed over the last three decades is that "word of mouth" communication gradually became digitized. Added to that, we digitized vast swathes of information that were previously the domain of books and newspapers and made them far more accessible. The effect of this is to bring the town square into the global arena; vastly increasing the amount of un-curated information that we must process. The victim here is reflection ⧖ **TIME,** and it is only through our ability to *manage* information effectively, and to build other tools that will help us maintain our perspective on it, that we can turn information overload into information abundance.

Scaling networks while compressing ⧖ **TIME** lies at the root of our inability to process information. We offer ourselves little time for reflection before responding. We have always had networks that moved fast. Consider for a moment how fast a rumor spreads throughout a high school. Technology now means that Twitter is our high school, but its reach is global. Just like rumors in high school, however, we can choose to believe what we see on Twitter critically or uncritically.

The other part of the network problem we face is one of frequency. Because the network is so much larger, there are many more stories being told and we are barraged with information from a plethora of sources. However, the same tools we used to manage rumors in high school (critical thinking and an evaluation of sources) can be employed to discriminate between the stories we need to worry about (and *how* to worry about them) and what we should file under background noise. To achieve this, we need to wrap our heads around the network itself as a tool.

John Naughton suggests one way of understanding this tool is to think of the network as an ecosystem.

> Looking at our media environment through an ecological filter leads us to a different kind of analysis. For example: It obliges us to treat it as a *system* distinguished by strong inter-relationships and dependencies between its constituent components. These components may look very different from one another, but they are inextricably bound together. The system is, as the old adage puts it, "greater than the sum of its parts." This means that analyses based on so-called "reductionist" studies of individual components taken in isolation are likely to be misleading. Any change is *systemic*: when one component changes, the effects ripple through the entire system. (Naughton 2012, p. 115, emphasis in original)

This is a useful way of thinking about the vast interconnected information sources that confront us these days. We need to look past the technology to its underlying purpose. Not that there is a master plan at work here, but, as Naughton suggests, what we are seeing is the organic growth, mutation, and adaptation of our communication pathways. It has to be thought of as a network of ideas first and understood in that way.

Networks are far older than networking technology and are profoundly human, not technological. The means have changed, but the connections are as old as human societies. They are part of what makes us human. Technology changes this but doesn't alter the fundamental reality that we created Facebook and Twitter, not Mark Zuckerberg, Ev Williams, and Jack Dorsey.

The network is a tool just like any other technology. Its purpose is the elimination of distance factors in our communication streams. We no longer require long-distance phone calls, physical mail, or travel to communicate at a distance. Posts, texts, or video chats have almost zero opportunity costs other than the ⌛ TIME needed to create them. We can now seamlessly broadcast our thoughts to many readers and viewers, even those we don't actually know. This reality is new, not the technology itself, and it is our ability to all be broadcasters that has reshaped our notions of ⌛ TIME.

In 1968, just before Douglas Engelbart did his "Mother of All Demos," Bob Taylor and J. C. R. Licklider (who funded Engelbart's work) gave a fascinating talk on "The Computer As Communication Device." (Taylor and Licklider 1968) Both Taylor and Licklider were trained as social scientists, and both saw the emerging technology in much more humanistic form than the technical virtuosos that dominated the field in the 1960s. Teletype machines and punch card readers defined most interactions with a computer during this period.

It was the interactive human networks facilitated by the computers that fascinated Taylor and Licklider. Both were of course aware of the work that Engelbart was doing at SRI. Taylor was overseeing a new project through his post at the Defense Department's Advanced Projects and Research Agency (ARPA): connecting computers into what was to become the Arpanet. In this talk, they put together these two strands and meditated on how computers might work if they augmented human intellect and simultaneously connected networks of augmented humans. They conclude their talk/paper by noting:

> *First, life will be happier for the on-line individual because the people with whom one interacts most strongly will be selected more by commonality of interests and goals than by accidents of proximity.* Second, communication will be more effective and productive, and therefore more enjoyable. Third, much communication and interaction will be with programs and programmed models, which will be (a) highly responsive, (b) supplementary to one's own capabilities, rather than competitive, and (c) capable of representing progressively more complex ideas without necessarily displaying all the levels of their structure at the same time-and which will therefore be both challenging and rewarding. And, fourth, there will be plenty of opportunity for everyone (who can afford a console) to find his calling, for the whole world of information, with all its fields and disciplines, will be open to him-with programs ready to guide him or to help him explore. (Taylor and Licklider 1968, p. 40, *emphasis added*)

Licklider and Taylor conceptualized their network with information exchange as its primary function, building on Bush's and Engelbart's vision of what was possible with computing technology. They foresaw it as a critical tool for curation and not a firehose of information. Limiting information flows was just as critical as establishing them, because information without a filter is just noise. A series of cartoons that illustrate these points accompanied their 1968 *Science and Technology* article.

Interactive communication consists of short spurts of dialog....

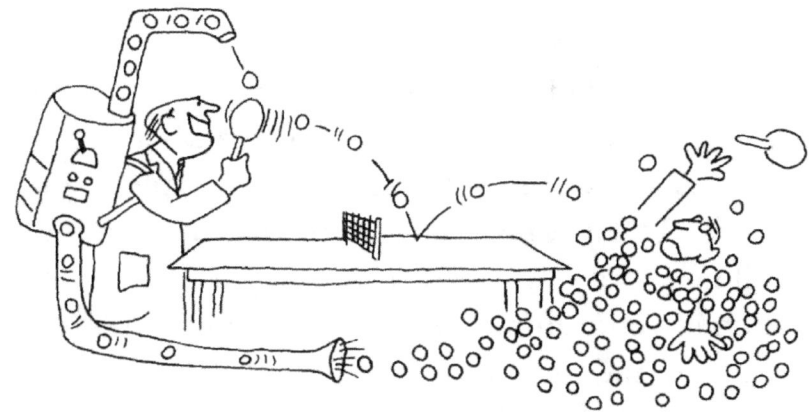

VERSUS .... filibustering destroys communication.
(Images from Licklider and Taylor 1968, pp. 34, 35)

Now imagine the ping pong "shooter" in the second image as the network (or Twitter) itself and you can see the fundamental problems we are confronting today. Our central challenge on the network is to scale and moderate our own conversations within it. To accomplish this, you must first establish a sense of the way it changes our stories. Just because people are talking doesn't mean communication is happening. The trick is turning the communication flow to your favor by filtering out the noise and curating the rest. Howard Rheingold suggests,

> Making informed decisions about where to deploy your attention begins with realizing that nobody can ever take advantage of all the interesting opportunities the Web presents us. Know that you have to say no, know what you are saying no to, and know why you are doing so. (Rheingold 2012, p. 246)

In the last chapter, we saw how our storytelling canvases expanded from the linear textual into the visual and beyond. Digital tools added more dimensions in ✪ **SPACE** for creative exploration. Digital ⧗ **TIME** opens up a whole new set of possibilities for the stories we create, both intentionally and unintentionally. To achieve this, we also have to understand the network itself as a story. Vannevar Bush's concept of breadcrumbs may provide one metaphor to understand the network, even if we can't always see them.

We constantly leave digital breadcrumbs in our wake. Sometimes, this is intentional when we tag ourselves as being somewhere. Sometimes it's unintentional as we buy products or take an Uber somewhere, leaving behind tokens about habits or location along the way.

One challenge inherent in the current environment is how to make these breadcrumbs transparent to those who leave them behind. They should be more valuable to us than they are to Facebook, Google, Amazon, Apple, or Uber. However, this can impede monetization if the product you are selling is access to our breadcrumbs. Minimizing the value of the individual to the system is a legacy of the Industrial Age, where humans were nothing more than productive and consuming widgets.

The fundamental problem with the network is not that it gives us too much information. It is that it cannot give us the right information with proper context. Hypertext, as developed by Tim Berners-Lee, models a stack of papers better than the ideas they contain. We have become trained to walk down linear pathways of information, whether that be by following a series of links or scrolling through pages and pages of chats, Tweets, or Facebook posts. These are linear information processes and often lead to blind alleys or, even worse, the donning of blinders. These kinds of informational streams are analogous to writing in text, as described in the previous chapter.

We work our way down an informational alleyway every day. The more adventurous of us may try a few of the doors along the way (hypertext links), but most of us struggle to see beyond the walls on either side. Again, channeling ideas in this manner benefits those who build the walls (and plaster them with advertising). It does not benefit the individual, nor does it benefit innovative thought. We never see the intersections that Johansson refers to, nor do we see the unexpected liquid networks described by Johnson. We are often totally out of control.

Traditional media has acted as society's information filter for decades. However, they are ill-suited for the curation task in the digital age. Linear, ⏳ TIME -based media are disconnected from the temporal flow of the digital age. Even 24-hour "news" outlets are subject to outdated storytelling conventions. They also skew toward the sensational in order to attract viewers. When they do engage with the network, they often do so clumsily at best, and maliciously at worst. In their minds, individual tweets from certain people themselves often create "news."

Mass media distorts information. It's like looking at a vast pile of apples, picking the extraordinary one, and then holding that one up as representative of the entire pile. Because of the speed of the news cycle, traditional news outlets, particularly "hot" media (as McLuhan describes it) such as television, lack the very perspective required for accurate contextualization of breaking events, much less explain the larger technological and societal forces pushing on the world today.

To regain control of our information flows, we need to deploy some of the tools we've been considering in previous chapters in the service of visualizing dynamic information, not just static concepts. Some of these kinds of tools are emerging to provide us with a sense of perspective to perceive networks of ideas. Text is unidimensional. Its consumption may allow for forks in the path but there are hard limits on how we can gain larger perspective except through the labor of reading a lot of diverse texts, a task few outside the Ivory Tower have the luxury of doing.

The speed of the network leaves us precious little time to explore in this way, so we easily lose sight of the forest for the trees. Visual thinking, which we explored in the last chapter, is another way of processing information and it, at least, gives us a second dimension to work with. We can enhance its power by adding a third, spatial dimension to its capabilities. As we have already discussed, understanding the network requires a way to process the fourth dimension, ⧖ **TIME**, as well. We do this imperfectly in our heads, but most of us find it difficult to hold four dimensions in place simultaneously, much less being able to share that or have it effectively shared with us.

When we attack the problem of understanding the network with more advanced tools, however, we comprehend the power that we can harness to contextualize ideas. One of the most striking instances of this is the Open Syllabus project and, in particular, its interactive visualization of over 6 million syllabi (**galaxy.opensyllabus.org**).

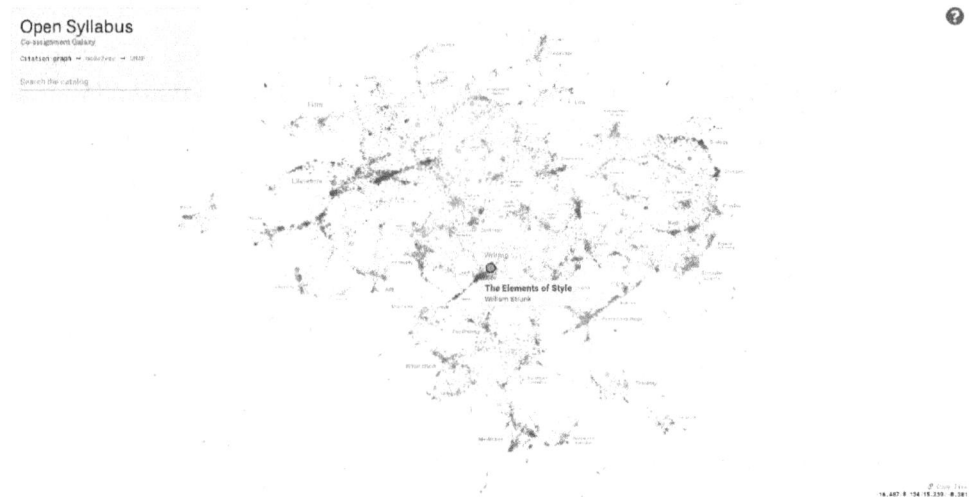

Open Syllabus Visualization

This visualization is an amazing tool for exploring the network of ideas that make up higher education. It is a representation of all the books and articles assigned as readings to students, grouped by numerical proximity. Those disciplines that cite the same resources clump together. Taken at a higher level of abstraction, this is a completely new way of looking at the information networks that make up the invisible threads of our education systems. It represents millions of data points in a clear, multidimensional way using both visual (a scatter diagram) and temporal (the ability to click through layers of data) tools to represent its ideas. This is an example of a tool that can help us grapple with the information overload that is the root of our struggles with the network in ways that textual media cannot do. Rich platforms track exchanges and open up doors to novel combinations of ideas.

This kind of tool requires access to the data necessary to create it. In order to capture the data, the Open Syllabus Project scraped it from thousands of university sites that post their syllabi. Not all platforms are that open in their sharing of data. Of the big, commercial platforms, Twitter is the best at sharing data points and there are several visualization platforms out there using its data to parse information in new ways. Researchers at North Carolina State University have developed a tool for measuring sentiment on Twitter that depends on their search application protocol interface (API).

Another avenue for commercializing tools is to show their utility by creating interesting visualizations. Maptimize, a Graphical Information Systems (GIS) developer, is using Twitter to showcase its software's capabilities to plot the geographic location and select hashtags of the last 1 million tweets (**onemilliontweet.com**). It then constructs heat maps and to show where the tweets are coming from.

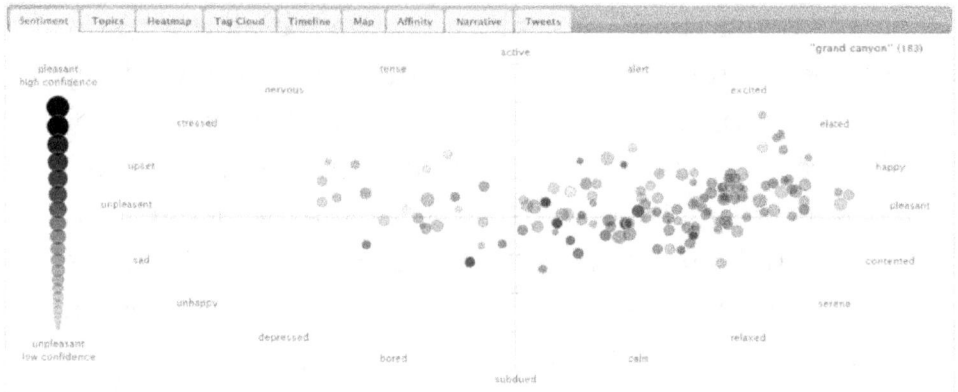

NC State Twitter Sentiment Visualization

These tools are just one way to visualize information, but I show them here primarily to show what is possible when we zoom out and see the network instead of the noise. As a social scientist, I am always concerned with sample sizes. Understanding what large numbers of people are doing and thinking is a useful filter for processing vast amounts of data. Understanding what Joe Smith of St. Louis is mad about today is a lot less useful. Twitter is also not broadly representative of the population. These tools measure outputs and 10% of Twitter users account for 80% of Tweets. This 10% likes to write about politics, skews female, and skews slightly liberal in the United States according to a 2019 study by Pew. (Wojick and Hughes 2019) Twitter only shows you a select part of the network. These limitations will therefore distort any analysis using Twitter data. We still have much work to do to make the network transparent to its users, but these kinds of tools represent a clear path forward.

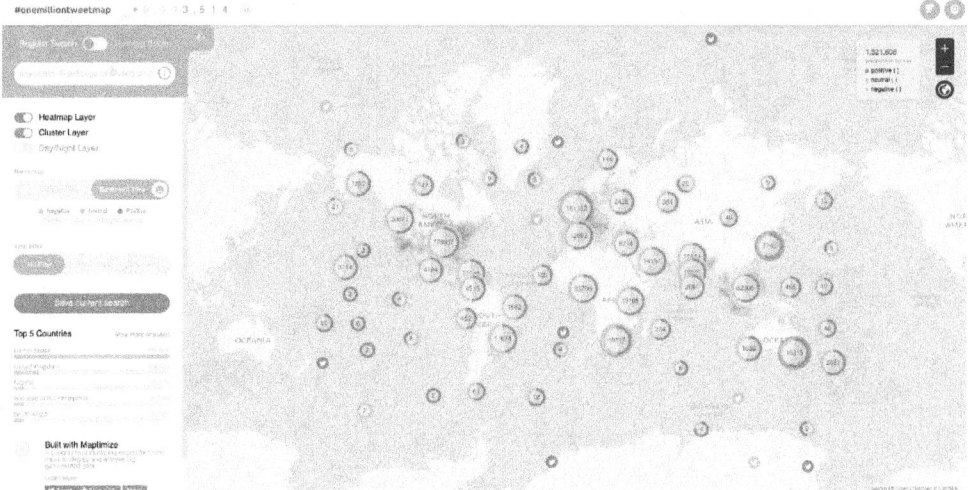
One Million Tweets Visualization

A recent study from Harvard University shows the possibilities of using visualizations to understand the broader news ecosystem. During the 2020 election campaign, a false narrative developed around voter fraud. A group of Harvard researchers scraped a wide range of media sources, including Twitter, to follow the path of the narrative and how it propagated throughout the information universe. Tracing the story with the aid of visual cues they could determine the frequency and timing of stories on the subject were not random but part of "an institutionalized rather than individualized disinformation campaign." (Benkler et al., 2020)

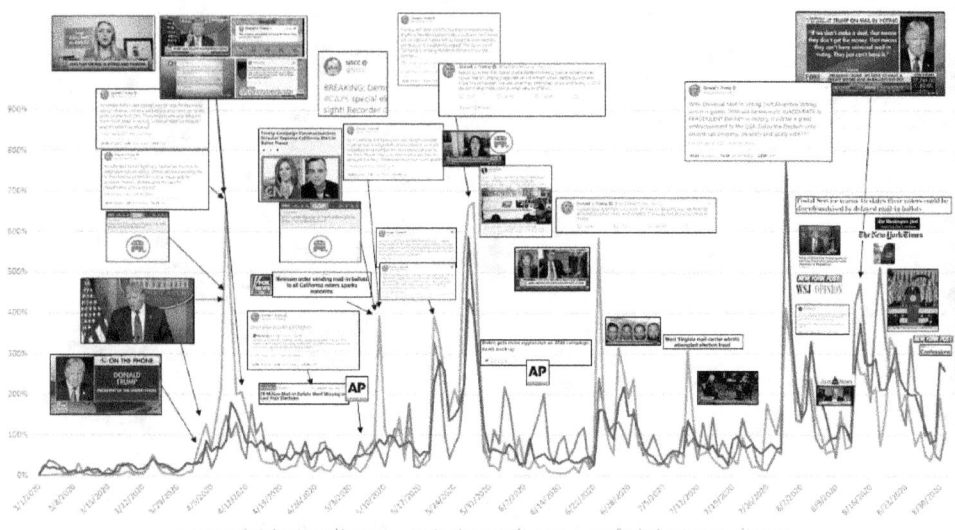

Visual Media Analysis of Voter Fraud from Harvard Study

The best of these visualization projects presents us with both a microscope to see what others are saying and a collective mirror of our habits. They help us gain perspective on the narratives that tear like wildfire through our information ecosystems.

We have built what the English philosopher Jeremy Bentham called The Panopticon. His concept, developed to improve the design of prisons, made it so that the guards could see all the prisoners from a central location. The prisoners would then become productive workers rather than simply residents in their cells. This is an apt metaphor for the network, but it is a 2-way panopticon. We are both prisoners and guards of our information. The only question is which role we elect to play.

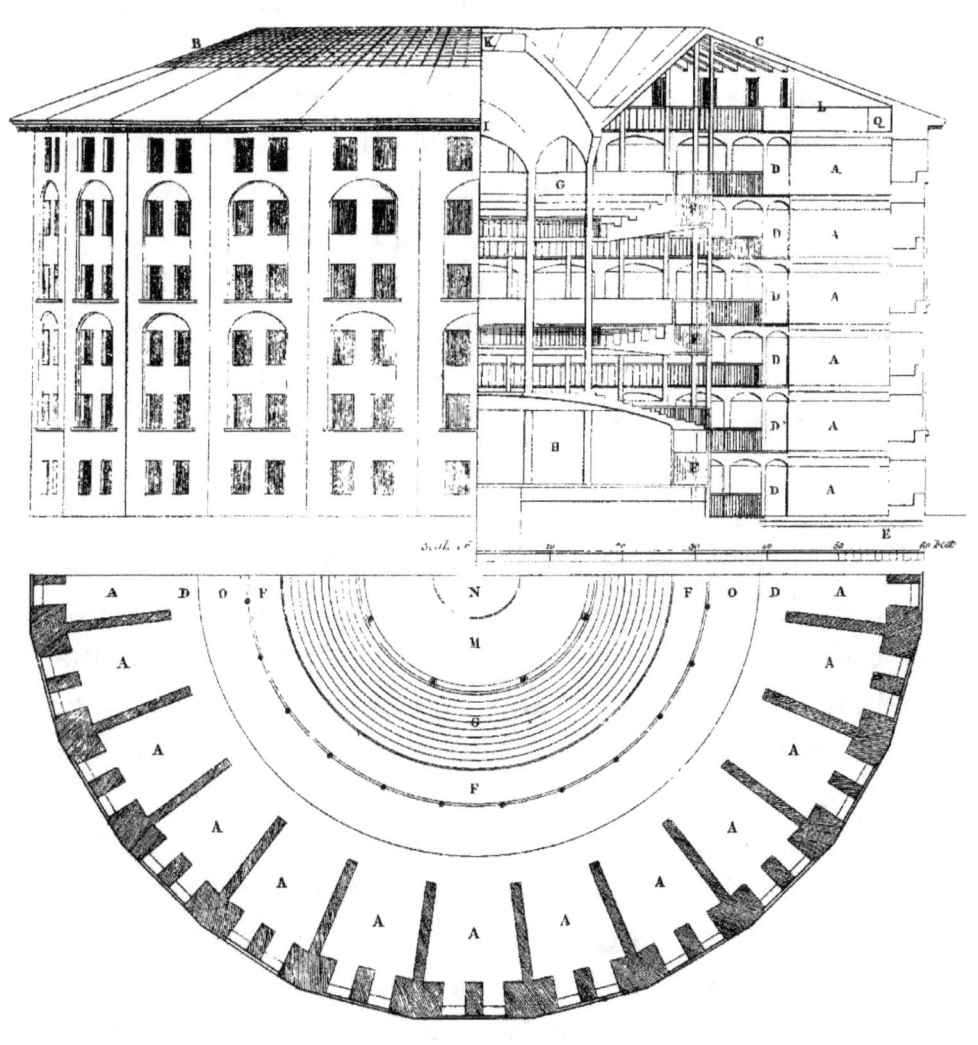

Jeremy Bentham's Panopticon Concept

We compete with those who seek to control our information for the role of guard. Getting the upper hand in our relationship to the network is key to whether or not we are the seers or the seen in the digital network. The challenge is that, unlike Twitter, the social media universe does not freely share its data.

This is because most social media companies depend on information asymmetry as a core element in their profit model. They sell data on customer behavior, so making that data available to those creating it doesn't make business sense. They also sell their ability to manipulate that behavior through careful micro-targeting of information through their understanding of our observed habits. Information asymmetry has the pernicious side effect of reducing the customers to rats in a technological maze, where the social media company can analyze and monetize their behavior. The system disempowers them from understanding any sort of context or perspective. This is industrial in scale and philosophy. It could come straight from the playbook of Edward Bernays, the progenitor of modern psychological marketing.

One simple solution to this kind of corporate behavior is not to patronize those entities that take your data curation out of your hands in favor of those that empower you. That's often easier said than done, as even data transparency has become the subject of marketing. The complexity of the systems also makes it difficult to track what's going on with your digital footprint. When consumer action fails, government regulation is often necessary. Transparency is an obvious place for government intervention. Instead of trying to restrict information flows, the aim of any regulation should be to remove barriers to its unfettered flow. Without transparency, curation of our information flows becomes much more difficult. Transparency should be at the core of any government regulation of social media technologies.

Transparency drives the positive effects of the digital network. Sunlight has fallen on areas that were previously hidden from view. To address issues ranging from systemic racism to societal inequity, it is essential that we shed light on a diverse set of issues. The network has made that possible in unprecedented ways. Instead of 1984 (or 1964, for that matter), the digital era has led to a diversity of stories being told that were largely ignored by the centralized narrative media outlets of the industrial era.

These stories were always there, but almost never surfaced in the dominant narratives. We have seen a wave of stories, from Black Lives Matter to Me Too to Occupy Wall Street, emerge as the network has given previously disenfranchised groups access to unfettered broadcast capabilities. There has also been a reactionary backlash using the same tools as the Alt Right movement and other groups seek to undermine those stories with their own versions of the past and present.

Our central challenge lies in using the technologies that have created this diversity to manage it effectively. Sophisticated tools that allow us to analyze and understand the root causes of the information flows are essential for separating legitimate grievances from those who seek to maintain power at the expense of the powerless. Redundancy promotes diversity. From its inception, redundancy was at the core of the design of the internet. To repeat John Gilmore's axiom, "The Net interprets censorship as damage and routes around it." (Elmer-Dewitt 1993)

There are also more difficult aspects to the changes that technology has created in our information flows. While transparency is critical to understanding the significant changes in persistence and discoverability that demand new perspective contexts, it also means we can never escape our past mistakes. The physical media upon which we inscribed it or the reliability of memory has always limited persistence of information. Now, once something is digital, we can endlessly duplicate it and search for it. This has resulted in the network biting back. Persistent, sometimes culturally inappropriate, information unexpectedly surfaces to ruin careers and disrupt trust. These are all products of the network. Toleration and reflection ⌛ **TIME** to establish context are essential for separating mistakes from ill intent.

Curation is absolutely necessary in this world, but we need much more sophisticated tools than those that were provided by Walter Cronkite in the 1960s and 70s to start with. Then we need to develop cultural standards that incorporate the perspective that those tools will grant us. This will take time.

The acceleration effect that the network has had on the speed of information flows throughout our societies has created distortions and eddies. We need to understand if these are ripples or tsunamis and how to contextualize them. While technology may be essential in facilitating our capacity to see what's going on, it will be our human cultural and organizational shifts that are once again critical to navigating competing narratives. Cultural shifts require reflection ⏳ **TIME** to percolate. Human societies are slow to change, no matter the circumstances.

In the last chapter, we discussed how the development of visual literacy is critical in a digital world. To this, I would add network literacy. This is the need to understand where your messages go as well as how the flows of various networks condition information that you receive. The flip side of this is to recognize the role we play in managing our information flows. These literacies require transparency from the network.

Transparency makes possible curation. Coupled with persistence; these two factors facilitate the building of trust. We have become numb to the constant barrage of messages sent to us by friends, commercial, and political interests that seek to sway us. The lines between these groups have become increasingly blurred. Trust has become the coin of the realm in a world of false appeals. In a 2019 Pew Survey of experts reflecting on the next 50 years of the Internet, Greg Shannon of Carnegie Mellon University remarked,

> "Trust will be a critical social asset. Those communities that value and promote trust will have more life, liberty and happiness. AI and IT will allow communities to ensure varying degrees of

> security, privacy, resiliency, and accountability in building trust. Being trustworthy *all the time* is stressful given that trust is based on competency, dependability, honesty, loyalty, boundaries and sincerity." (Stansberry 2019, p. 114)

The lack of transparency in how our digital footprints are being used is a violation of trust. We trust network-based companies to take good care of our data. This is often a misplaced trust. We trust in the news media to unearth significance from the vast deluge of events that are now made visible by the network. Having that information misrepresented in the name of their own profit instead of the public good is also a violation of trust. Nigerian princes rarely leave their fortunes to complete strangers, but again, these kinds of cons involve the brokering of trust in a medium unfamiliar to many. Phishing attacks involve trust, as do most forms of technological exploitation (as distinguished from hacking) we get in email and through social media.

We make ourselves vulnerable to these trust violations when we try to process Digital Age information flows using Industrial Age tools. The pathway for mailing a letter is fairly linear and highly regulated. Most people don't realize that the pathway an email follows is far from linear and almost completely unregulated. While both pathways are equally obscured, it is possible to make the electronic path instantly visible to both sender and recipient. The information is already there. It's just a technical exercise to parse it. This information could be simplified and extended to other forms of electronic communications like chats. This would make it more transparent to the casual user who is more concerned about getting a message through than where it went in the interim. However, these pathways are essential for trust.

Linear, industrial information processing tools do a poor job of filtering out the trustworthy from the untrustworthy in a Digital Age information environment. That is why we need to understand the *networks* of information that link our stories with one another. Trust requires the conveying of the legitimacy of the sender of information. We depended on hierarchies of power in media, business, or the government to convey whether a sender was legitimate. The Digital Age works against these ordered flows because its flows are impossible to apprehend. They no longer have to pass through a printing press or broadcast studio. Digital information flows essentially give us three choices:

1) Will we continue to be dominated by elites that determine the trustworthiness of information?
2) Do we simply trust anyone with access to the network?
3) Do we evaluate legitimacy ourselves with effective analysis tools?

The historian Niall Ferguson provides one view in his analysis of the competition between networks and hierarchies for control of information. In *The Square and the Tower*, he argues that networks have disrupted power hierarchies, but power has always reasserted itself. This is because, in his view, hierarchies are essential to avoid anarchy. He is deeply critical of the notion that flat structures can be stable and will persist. In the end, there is only anarchy or hierarchy.

> We, too, must look back longer and ask ourselves the question: is our age likely to repeat the experience of the period after 1500, when the printing revolution unleashed wave after wave of revolution? Will the new networks liberate us from the shackles of the administrative state as the revolutionary networks of the sixteenth, seventeenth, and eighteenth centuries freed our ancestors from the shackles of spiritual and temporal hierarchy?....
>
> 'I thought once everybody could speak freely and exchange information and ideas, the world is automatically going to be a better place,' said Evan Williams, one of the co-founders of Twitter in May 2017. 'I was wrong about that.' The lesson of history is that trusting in networks to run the world is a recipe for anarchy: at best, power ends up in the hands of the Illuminati, but more likely it ends up in the hands of the Jacobins. Some today are tempted to give at least 'two cheers for anarchism'. Those who lived through the wars of the 1790s and 1800s learned an important lesson that we would do well to re-learn: unless one wishes to reap one revolutionary whirlwind after another, it is better to

> impose some kind of hierarchical order on the world and to give it some legitimacy. (Ferguson 2014, pp. 549-550; 553-554)

Ferguson argues hierarchies will inevitably re-emerge. He assumes that anarchy and hierarchy are the only two options. I would argue that the digital information environment gives us the tools to consider other conceptions of order. Hierarchies will re-emerge only to be undermined repeatedly by the digital world that Claude Shannon imagined in 1948, but the options are richer than a bipolar choice between anarchy and hierarchy.

Hierarchies survive by controlling information. Information in the Digital Age is uncontrollable. The cycle of hierarchy —> disruption —> anarchy —> new hierarchy that Ferguson describes, like everything else, will be accelerated by the speed with which information travels. The speed of digital oscillations between hierarchy and anarchy may serve to dampen the pendular effect as we process new realities (like we process new technologies) on an increasingly rapid basis.

Like Ev Williams, I used to believe in the utopian transparency theory of the internet: that free information will lead to an era of greater democracy and freedom in the world. I still think this will happen, but strategic employment of digital tools will be necessary to avoid the pendulum that Ferguson describes. Hierarchies will always attempt to reassert themselves, but will find themselves repeatedly undermined in today's information environment.

We can only hope that fear of anarchy will drive transparency efforts, both private and public, toward a sustainable equilibrium that empowers humans, not the systems that seek to control them. It's really a challenge of understanding, leveraging, and augmenting the realities of shifting ecologies of information. The network will disrupt tyrannies. It will create them as well, but only temporarily. The modulation between the two extremes will dominate our lifetimes. This is true of government as well as business, as both kinds of institutions struggle to find a new "normal" in a world characterized by uncontrollable flows of information and communication.

Understanding the oscillations of the network is critical to streamlining any organization for the myriad opportunities and challenges these new realities create. This is not a technical exercise. It is a social one. For instance, a new role for government regulation might be to require corporations to exercise a measure of transparency when they collect digital tokens from us, as well as the algorithms being used to process them. I am confident that, like all digital information, this will become available. The only question is whether that occurs as part of an orderly process of regulation and enlightened business practices or through a disorderly one of hacking (or, more likely, a combination of both).

None of the issues we see with the network of human communication are new. Like many "technological" issues, they were largely buried by the anomaly that was the industrial era, where technology restricted mass communication to those who controlled the press and, eventually, the airwaves. "Broadcast" technology started with the automated press in the 1850s and was electrified by the radio in the 1920s and television in the 1950s. It is a fallacy to think that humans did not engage in intensive communication before 1850. Instead, communication was profoundly local, restricted to the village or town in which most lived.

Our disconnects happen because we've adapted to what was actually an anomaly in human history: the disruptive effects of industrial communications systems. Tom Standage made this point clear in *The Writing on the Wall*, when he went through and described the startling parallels between the affordances of the digital social media world and the pre-industrial means through which humans interacted with one another.

> New Media is very different from old media but has much in common with "really old" media. The intervening old-media era was a temporary state of affairs, rather than the natural order of things. After this brief interlude — what might be called a mass-media parenthesis — media is now returning to its preindustrial form.... [T]he historical forms of social media have enough in common with the modern

> sort — in their underlying social mechanisms, the reactions they provoked, and the impact they had on society — that they can help us reassess social media today, and the contemporary debates it has triggered. (Standage 2013, pp. 240-241)

Once again, we see how the industrial world created a disruption to our humanity. The broadcast revolution coincided with the urbanization of the population. Urban culture is the social construction most people alive today, especially in the developed world, are most familiar with.

Privacy was a priority in the industrial environment because of the overwhelming number of people concentrated in a relatively small area. Trust became transactional. In the industrial world, we learned to trust institutions, not individuals. The digital network has eroded this norm.

Trust was an important currency in the pre-industrial village. Life in the big cities atomized it. Trust and communication are intimately and intricately related. There are few secrets in a small town. There are many secrets in the big city. Broadcast technology, from the broadsheet to Fox News, effectively depersonalized news because it assumed that everyone needed the same information. It was technically and philosophically remote from the individual.

The advertising industry that developed in parallel to broadcast media also played on this kind of institutional trust. We even depersonalized our heroes as they shifted from our family to remote superheroes. Communication is at its heart about finding personal meaning. That is why people get so excited about seeing themselves on television. It also explains the success of YouTube, SnapChat, and other video sharing services. Even if your audience is tiny, it's still bigger than one.

Many of the problems we have with inappropriate sharing stem from this clash of cultures between the intimate and the impersonal. We still instinctually believe that communication operates at the level of the village pub. We take media narratives of all kinds as pronouncements from on high and follow the peccadillos of celebrities as we once followed the eccentricities of the king or the lord. These often have little bearing on our day-to-day existences, but we have a tendency to make them a central part of our identities. This is not new. What has changed is that the digital network has made us all into potential celebrities.

The village pub is now a global network of conversations. However, what we say on Twitter does not stay on Twitter. In extreme cases, it migrates to the traditional media, which increasingly depends on tweets for raw fodder. Actors give and then break trust repeatedly. All too often we think some of those actors are our close friends, only to find out that many of them are the modern equivalent of the flimflam man.

The level of noise on the network obscures the incredible opportunity it holds for us personally, professional, and as a society. It used to be that we had to go to places like Harvard, the Sorbonne, or Oxford to have the wide-ranging interdisciplinary conversations that are now possible remotely. For both good and ill, we have realized Licklider and Taylor's vision of interacting with people "selected more by commonality of interests and goals than by accidents of proximity." I regularly interact with friends and colleagues to exchange ideas with little or no mind taken to geographical proximity. This book itself would not have been possible without such interactions.

The network gives us the ability to more easily cross disciplinary boundaries, a kind of global *Medici Effect* that would have been difficult in the era of unidirectional networks that existed before the turn of the last century. Sure, the elites of those areas conversed extensively by letter and the scholarly press. Now, however, those conversations have become massively more diverse. Brilliant ideas no longer need to be confined to a small intellectual elite lucky enough to have received the requisite education to allow them to inhabit elite institutions or the social and economic standing necessary to gain admittance to the halls of power. Now, political, economic, social, and scientific thinkers can theoretically communicate from anywhere and at any time.

This applies equally to the business sector. Within the firm, we do not need to confine decisions to the boardroom. They can be widely discussed, iterated, and collaborated upon by a diverse set of actors within the organization. These actors can be geographically dispersed and we can adapt the same tools that are being used to create persistent social media for collaborative, organizational purposes. It amazes me how many organizations cannot grasp the fundamental opportunity these affordances provide. We still structure most firms for hierarchical information flows that are a product of the industrial mindset. This is now increasingly a liability as firms cannot leverage their internal talent effectively.

> The culture of creative disobedience that draws innovators to Silicon Valley and the Media Lab is deeply threatening to hierarchical managers and many organizations. However, they are the ones who most need to embrace it if they are to support their most creative workers and survive the coming age of disruption. Innovators who embody the principle of disobedience over compliance do not only increase their own creativity — they also inspire others to excellence. (Ito and Howe 2016, pp. 140-41).

I have been in many meetings where industrial-minded leaders were decrying the propensity of "millennials" to go outside the accepted channels of communication with clients. This often threatens the structured industrial hierarchy built into projects. The network, however, is biased toward free-flowing information and younger employees, having grown up in that kind of information environment, are unlikely to see problems with flattened communication strategies.

I dislike the term "millennial" because I think it masks a profound change in attitude behind a stereotype. Anyone born in the developed world after 1990 is essentially a child of the network. Millennials are used to the disruptive flows of information that wash around the world. They have grown up recognizing the creative potential of distributed creativity on platforms such as YouTube. Their world view is pliable. The industrial world view is not.

Fluid networks also demand that we rethink our notions of privacy. During the Industrial Age, we learned to hide behind the social, hierarchical structures we constructed to protect us from the cultural dislocations of moving the vast majority of humanity from trusting villages to huge, impersonal cities. This often caused the establishment of rigid networks of trust to protect our privacy. The digital network is profoundly destabilizing to these structures of privacy. It has become increasingly difficult for humans to hide in the new/old environment of the globally networked social media village.

Cultural backlash is inevitable. Power structures protect their privileges, whether we are talking about implicitly privileged groups or explicit political actors within a given society. Again, this is a human problem, not a technological one. The fundamental nature of the industrial, hierarchical organization goes against the technological logic of the network. Systems will always attempt to foreclose possibilities long before they will exploit opportunities. The business press may extoll the onset of the "new economy." However, most firms are stuck firmly in the old industrial communication and cultural paradigm. Undermining systemic culture is always a difficult proposition, no matter how strong the logic imposed by the technology might be (we will explore this in greater detail in the next chapter).

There are some fundamental principles that govern the paradigm that the networked world imposes on us. First, we must accept the reality that any information that is digitized (and all information is digitizable) is now infinitely remixable and shareable. Second, we must accept the fact that we can no longer control how that information is remixed and shared.

Information used to be the scarce commodity in our ecologies of knowledge. This has changed. Information is more plentiful than ever. Vannevar Bush's breadcrumbs are now the scarcest resources. The real challenge in figuring out the network is to *trace* information, especially information that you find useful (curation), enlightening or infuriating, back to its original source (transparency) and then contextualize it (reflection). Transparency and curation are the only way that we can ever hope to separate the wheat from the chaff in the modern information environment. On the positive side, however, information often resembles an onion that needs peeling. Unlike an onion, which can lead to tears, deconstructing information can lead to unexpected vistas and insights.

Perspective is everything in the network. First, we have to be mindful of our storytelling and how it rebounds within the network. It used to be the mistaken group email that did us in there. Now it's an ill-advised Tweet we posted 7 years ago. Whether or not we like it, we all live in public. The sooner we accept that reality and adapt to it, the more we can leverage the network for our own creative pursuits effectively. Second, we have to understand that everyone else is trying and (to a greater or lesser extent) failing in the same activity as we are. This is the stream we have to contend with.

Those who build mechanisms of reflection rather than reaction into their systems have an advantage in this environment. We have to grasp ⧗ **TIME** and let humanity drive the network, not the other way around. Instead, we need to slow network speeds down to human speeds. We can gain a better perspective using tools like those described in this chapter to filter and process network inputs, such as the Harvard study cited earlier. (Benkler et al., 2020) The media should adapt to covering this kind of story rather than ones fed to it by Twitter.

However, as individuals and organizations, we need to recognize that aggregated information is often more valuable than anecdotal inputs and filter our reactivity accordingly. This also helps us assess the validity of information more systematically. The network exposes flaws in our understanding of ourselves and fallacies, such as appeals to authority and appeals to ignorance. This has more to do with our natural reactions to information overload (coupled with the speed of information flows) than any inherent flaw in the network. We need to learn to look for patterns of information in the same ways our ancestors did when they surveyed the forest for signs of predators and prey.

In the absence of tools for visualizing patterns of information, we must also recognize that slowing things down to human speed is our prerogative. Give yourself and your team ⧖ **TIME** to reflect and gain perspective on the information flowing around you. It is easy to contribute to the noise by reacting to it. Assuming a certain level of transparency, which I predict will become easier as tools catch up with information flow, more data will create clearer pictures of the shape and intention of those flows. Once we master the flows, our next challenge will be to develop structures that act as tools for turning information into innovation. That is the subject of the next chapter.

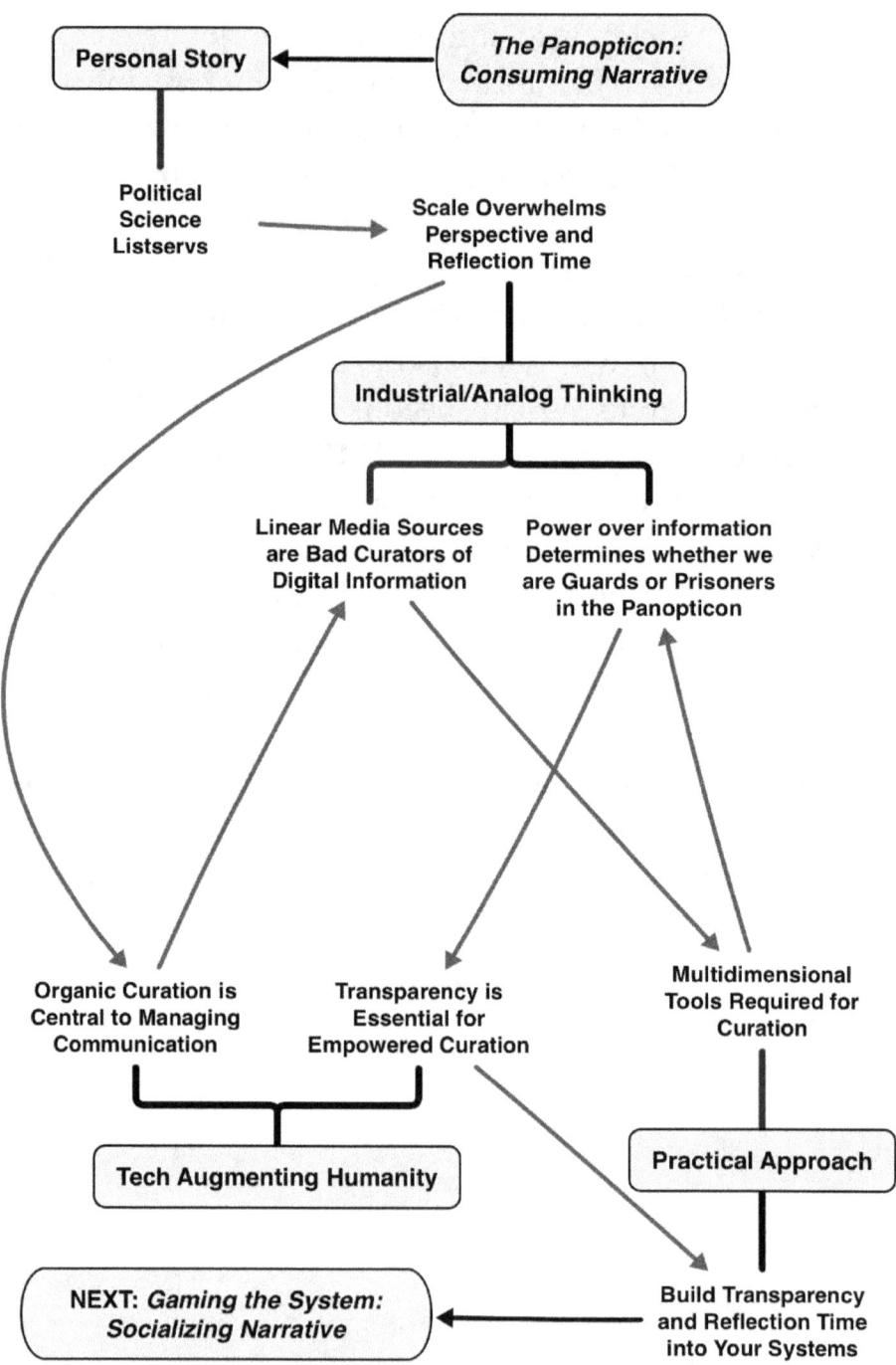

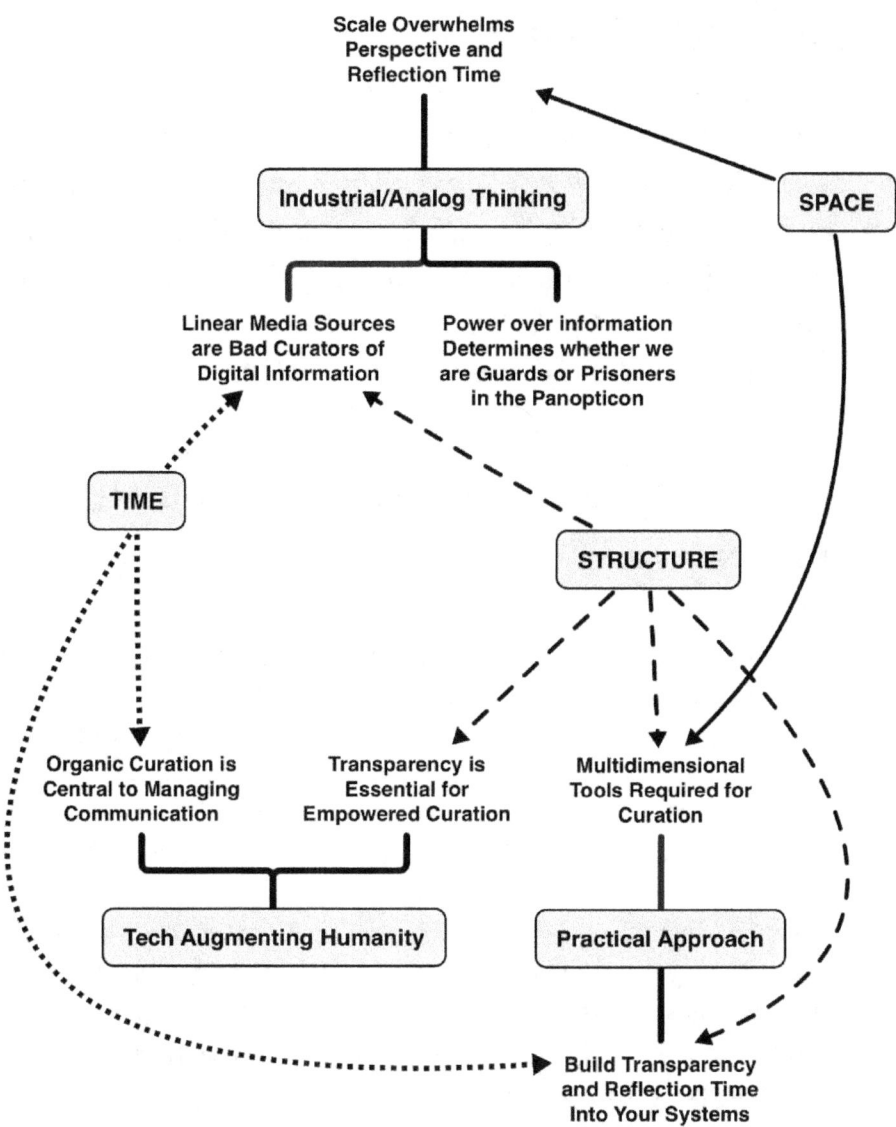

❈ STRUCTURE

Putting the stories we have captured and shared into action requires their application in practical (or impractical) ways. The Digital Age allows unprecedented amounts of remixing and creative activity. Creating structures of play are made possible by the vastly reduced opportunity costs afforded by the technologies of the Digital Age. Leveraging structures of play are central to producing innovation.

2.3 Gaming the System: Socializing Narrative

We all know that Art is not truth. Art is a lie that makes us realize truth, at least the truth that is given us to understand.
Pablo Picasso, 1923

The computer is simply an instrument whose music is ideas.
Alan Kay, 2003

It's the people who think that things are open-ended, that things can still be changed through thought, through creativity—those are the true optimists.... Also, to me, a sense of the world being open-ended is absolutely core to being a good scientist, a good technologist, a good writer, a good artist, or just a good human being.
Jaron Lanier, 2018

It was the summer of 1978. I was the youngest camper at a summer camp in West Texas. One day, as a camp activity, we played this game called *Dungeons & Dragons*. The game immediately engaged me more deeply than any other game I've ever encountered. It empowered me to walk through my imagination. Upon my return home, I quickly bought the rule books and learned how to create worlds of my own. With relatively few exceptions, my only player was my younger brother, but I spent endless hours trying to create adventures that were both fun and made sense.

Looking back, I now believe that this is where my fascination with systems thinking began. In order to create adventures, you had to consider a wide variety of variables. The system could not be too difficult, or the player would lose interest. For the same reason, the challenge could not be too easy. Successful *Dungeons & Dragons* worlds contain many interlocking pieces. I quickly discovered that the more time I spent figuring out how these pieces work together as a system, the more immersive the adventure would become.

As an adult, I have continued to take part in active gaming. Although the D&D world occasionally tempts me, my interests have drifted to wargames. In particular, I have spent endless hours working with good friends, designing and developing better and better versions of the games that we already play. The same principles apply as they did in *Dungeons & Dragons*. Games construct narratives as we play them. I dislike games with preordained outcomes. The element of surprise is crucial to any gaming experience. The possibility of surprise is also critical to any form of experimentation. My friends and I have occasionally discussed whether we were wasting our time playing and designing wargames. I have always argued that we have become better organizational leaders and teachers through the refinement of our rules to balance expected with unexpected outcomes.

Games are everywhere. Teachers construct games when they plan out their classes. Any good institutional design incorporates an element of game design. Practicing scenarios in the safety of a game allows us to experiment with unanticipated outcomes.

This book is about creating and designing tools that augment our humanity. Games are meta-tools. We can design them to construct and reimagine other tools. Constructing narratives by designing and playing games is every bit as much a creative process as constructing narratives through graphics, text, or schematics.

Playing games is a tool for exploring complex challenges. They are self-constructing stories that incorporate the possibility of failure. A well-constructed game becomes a playground for ideas. In very simple terms, those can be ideas about how to combine and recombine cards to achieve the greatest potential value. Poker is essentially math, with a human variable thrown in. Wargames are exercises in probabilities with even the best-laid plans overturned by unpredictable rolls of the dice. The same things can happen in *Dungeons & Dragons*. It's partially the unpredictability of games that make them so entertaining. We can take risks with very little real-life costs other than our ⌛ TIME.

This can be quite an immersive experience if structured properly. Open-ended gaming creates an infinite canvas for play and exploration, which is central to engaging and immersing an audience in the experience. As Bryan Alexander points out:

> Immersion must also persist over time, iteratively. A mastered and unchanging game becomes a mere exercise, and an explored space bears little further exploration. Characters that do not change are often derided as flat and undeveloped; the desire for character development assumes an iterative arc. Therefore, successful immersion must progress over time, repeatedly establishing [Janet] Murray's sense of enchantment and engagement. (Alexander 2017, p. 100)

Immersion leads to a desire to play and explore and that can lead to Czikszentmihalyi's *Flow* state. (Csikszentmihalyi 1990, 2008) Game designer Jenova Chen has actively applied Czikszentmihalyi's Flow ideas to encouraging open-ended problem-solving through his concept of Dynamic Difficulty Adjustment (DDA). (Chen 2007) In Chen's model, he proposes an unbounded game where challenges have no maximum score, should provide immediate feedback, and should allow players to navigate tests strategically. What he is suggesting is an unbounded game without a defined end, much like Dungeons and Dragons.

Transitioning to Digital Age thinking, untangling global climate change, politics, and education are all unbounded games. We can use open-ended gaming's combination of play and narrative creation to turn open-ended gaming into a set of ideation tools that can bring people together to engage in the unbounded exploration of our challenges collectively.

To do this, we have to imagine games differently. Games we played as children like *Monopoly* and *The Game of Life* are all significantly bounded universes. Even Chess falls into this category. There is a finite set of moves and outcomes. The rules drive players down specific goal-driven paths. Unbounded games do not have clearly defined and fixed outcomes. *Dungeons & Dragons* may resemble a goal-directed quest, but those goals change over time and there is never any fixed end unless the players decide to define one. Bounded games have their use. Their constraints can often lead to considerable creativity, but as design exercises they can be quite limited unless they lead to broader conversations, including the iteration of the rules of the game itself. Of course, when we revise the rules, you've effectively turned a bounded game into an unbounded one.

We discussed extensively in Chapter 2.1 the possibilities of the infinite canvases created by the Digital Age to explore challenges in fresh ways. We can build on our individual canvases by using them to build games that allow others to join our explorations. Immersive, open-ended games have the potential to lead people into explorations that they might otherwise be reluctant to pursue. They can also introduce a group element which adds a collaborative aspect to both the problem-advancement and the journey to get there. Games create and expand upon the kinds of narrative possibilities that we have explored in the previous two chapters and are likewise facilitated by both digital tools and the digital mindset. Digital tools have allowed me to create visualizations that have sparked the creation and adaptation of games. These can turn the passive consideration of ideas outlined in Chapter 2.1 into active, cooperative exercises in modeling complex issues. Visualization has led me into constructing games that model complex issues in a way where the participant actively engages with challenges that may be difficult or even impossible to simulate otherwise.

These are journeys of the mind. The form of the game is secondary to the purpose. What matters here is how the game drives the journey of the mind. Given the state of computer programming tools, there are still significant technical barriers to creating open-ended games online. The idea here is not to suggest that most of us can create *Call of Duty*-type games. This kind of storytelling requires complex skills and planning, on top of the technical skills required to create a functioning virtual world.

However, much of the difficulty there (given the tools of 2020) lies in fixing the imagination in a dynamic environment. You can achieve the same, if not more profound, effects by *not* fixing the mind and instead giving the imagination a playground to play in. In this manner, game design is analogous to many other kinds of storytelling. You can either make a movie, fixing a range of characters and environments in the mind through the personification of actors and the artistry of the director and crew or you can let your mind do that by reading a book. Games are the same way. As I have extensively discussed throughout this book, the Digital Age has the power to extend our imaginative skills. So, for instance, when I played *Dungeons and Dragons* in 1978, I was physically in a room with all the other players. My son plays *Dungeons and Dragons* using online platforms that connect players across the country. His Dungeon Master is a thousand miles away.

In this chapter, therefore, we will not focus on the technical details of creating complex environments online (although that is a fascinating topic itself). Instead, we will focus on the power of the game to create imaginative environments that can then model scenarios or merely to prepare the mind for the mental exercise of working through challenges when they appear for real. This is another example of where strategically employed tools from the Digital Age can augment us as humans. Playing *D&D* today, I would not be forced to rely on my brother as the sole player in my constructed environments. Now I can extend those environments throughout the network and find players where they are. I can also use Digital Age tools to more easily design board and card games to stimulate both serious and recreational play.

As a consultant, I often use games as a way of stimulating outcomes and providing alternative ideas and brainstorming exercises. Using theoretical frameworks and designing game rules around them is much the same exercise as using historical frameworks and designing wargame rules around them. The digital age allows us to play with much less friction and resource requirements because, like all narrative structures, the digital equivalent of games is much easier to create and manufacture than in the analog era. Gaming leads to experimentation, iteration, and the fundamentals of design. These are essential skills in a digital environment.

Play is the creation of stories. Stories have always been a precursor to understanding. The Digital Age is all about lowering the barriers to iteration and experimentation. Like almost everything we have discussed in this book, play and design are old concepts, but ones that the Digital Age has democratized. By democratizing the tools of play, we have also opened up alternative possibilities for exploration.

Systems and paradigms are forms of games. They are sets of rules that everyone within them largely takes for granted. These can range from cultural norms to checklists. My classes are games. The system in which the students operate is also a game. One of our key challenges in the Digital Age, as described in the very first chapter, is to change the rules of the industrial games we play and make them more consistent with the Digital Age realities that we face daily. In previous chapters, we have talked about how we can use tools to manage ideas in both ✪ **SPACE** and ⧗ **TIME**. Those tools allow us to play with ideas to model pathways to the future. Open-ended play allows us to model these pathways collectively. Many of the issues we face are systemic. Systems are complex games. However, like hacking, games can disrupt and change systems themselves. Playing with systems is still play. Our digital tools give us new power in this arena.

I have been asking you to play from the very beginning of this book. It was play that led me into computing and it is play that keeps me innovating. Humans play when they open themselves up to technologies that augment their intellect. They play when they hack technology to achieve those ends. They play when they design environments for experimentation (aka, play). They play when they tell stories. They play when they use the network to spread those stories. Play is everywhere. Our greatest mistake has been to separate it from "work" and denigrate it as secondary to "serious" pursuits. Even when we play for the sake of playing, it can have significant consequences merely by increasing the "wonder" of intellectual exploration. Samuel Johnson recognized this over 200 years ago.

> It may sometimes happen that the greatest efforts of ingenuity have been exerted in trifles; yet the same principles and expedients may be applied to more valuable purposes, and the movements, which put into action machines of no use but to raise the wonder of ignorance, may be employed to drain fens, *or* manufacture metals, to assist the architect, or preserve the sailor. (Samuel Johnson 1825, p. 146)

Play was considered a luxury during Johnson's time. Indeed, this fallacy has persisted to the present day. People consistently overlook the reality that effective "work" is a direct product of "play." In 1960, Claude Shannon, who has figured into this story at multiple points, along with physicist Edward Thorp, spent the better part of a year playing with the crazy idea that they could defeat a roulette wheel using physics and math. In the end, they ended up creating a portable computer coupled with toe sensors to track the speed of the roulette ball around the wheel, and consistently beat the house odds.

This frivolous pursuit (and neither ever really wanted to cheat the house, it just presented a technical challenge to them) resulted in a burst of creativity. As Steven Johnson relates, "the mix of technology Thorp and Shannon assembled to crack the roulette game had not only never been seen before, it had barely been imagined…. Fifty years later, Thorp and Shannon's roulette wheel hack would become the most ubiquitous form of computing on the planet: a small digital device in your pocket, attached to headphones, with sensors recording your body's movements." (Steven Johnson 2016, p. 226)

Shannon, besides his mathematical insights, was a notorious tinkerer and created many devices by "playing" around with stuff in his basement workshop. As a graduate of Bell Labs, he had access to technology not available to most, such as the surplus transistors he used to build his pocket computer. He had, by some estimates, half-a-million dollars' worth of equipment (in current money) in his basement and constructed a wide range of "toys" with it including a mechanical mouse that found its way through a maze and other forms of "mechanical Turks."

Shannon had the advantage of being relatively well off, due to savvy stock market investments. He had access to the latest technologies through his association with Bell Labs and MIT. This combination of factors allowed him both the freedom and resources with which to pursue his playful passions. As the roulette example amply shows, however, that eccentric playing often led to real insight and a breaking of paradigmatic thinking about a range of subjects, from mathematics to artificial intelligence to computing.

Shannon is just one illustration of the close connection between play and innovation. Mihaly Czikszentmihalyi's concept of "flow" also describes what amounts to a state of play. Flow is a state where you are suspended between anxiety and boredom, achieving a state of intense productivity and creativity. In a well-constructed game, we also lose a sense of ⧗ **TIME** and enter a simulated environment. This helps us safely examine the challenges that lie in front of us.

It is clear that Shannon frequently enjoyed this kind of state as he tinkered with his paradigms, whether those were in mathematics or mechanics. I have seen this state in many others, however, most notably in young children. In the late 1960s, George Land started giving children a test of "divergent thinking" that asked participants to imagine as many uses for a paperclip as they could. Most adults came up with 10-15 results, most kindergartners came up with 200. These numbers converged with the adult number when students Land and Jarman tested them again in middle and high school. They speculated that this evidence suggests that schools educated divergent thinking out of students. (Land and Jarman 1992)

However, it is also quite plausible to assume that younger children are more able to play and therefore enter a "flow" state and that school disciplines us out of this over ⌛ **TIME** by creating a set of artificial strictures to be adhered to. Stepping outside of those paradigms is often punished by "failure," but it is only through stepping out of them that we can truly change the world. Shannon changed the world in part because he never stopped playing. Czikszentmihalyi's playful concept of Flow, taken a step further, extends to the ability to innovate and imagine worlds that do not conform to our current paradigms.

> The point is that playing with ideas is extremely exhilarating. Not only philosophy but the emergence of new scientific ideas is fueled by the enjoyment one obtains from creating a new way to describe reality. (Csikszentmihalyi 2008, p. 127)

Even once we get past the societal barriers, however, play was much more difficult in the Industrial Age because of technological considerations. Consider for a moment the tools available before digitization and how those discouraged experimentation by all but the intellectual elite (supported by universities or foundations), artistically exceptional, or wealthy (Shannon was all three in his later years). Translating your creative impulses is hard in a world dominated by text, fine art, and music. The barriers to entry are very high. Typewriters are relatively easy to operate, but writing is definitely not. "Playing" with text on a typewriter is extremely hard because once it's on paper, you can't easily reshape it without the use of scissors and copious amounts of adhesive — and that only works once. I have significantly played with the contents and ✱ **STRUCTURE** of this paragraph alone at least half a dozen times, with nothing more than a series of keystrokes.

Text was relatively easy to manipulate when compared to the kinds of audible or visual creation we discussed in Chapter 2.1. However, it requires a level of abstract thinking that is a difficult process to master. Technology can be a barrier to expression in the same way as musical instruments. As Alan Kay makes the analogy, "for most people, the piano has been the biggest thing to turn people away from music for the rest of their lives." (Kay 1986 at 44:22) Non-textual, visual expression required the mastery of paint, chemical photography, or motion pictures. As we've discussed extensively to this point, these barriers have dropped significantly in the Digital Age.

As a photographer, I have long struggled to achieve Flow in my work and make it "play." I'm doing my best work when the camera becomes invisible in my hands and I can focus fully on the moment of visual storytelling I'm trying to achieve. I have done this in an analog world using film and darkrooms. The technological struggle at that time required the development of significant skills in order to master the process from capturing the moment to reproducing it on film or paper. I once spent an entire year in the darkroom trying to master that aspect alone. Audio and film recording, using physical tape, mimicked this level of difficulty and was inaccessible to all but the few who had the time and money to pursue them. Achieving Flow usually wasn't the problem, technology was. This, mitigated to a small degree by the new technology of analog video recording, was the state of affairs that existed as recently as the 1990s.

The 25 years since mirror the evolution I experienced with my own technological tools. During that time, I've moved from working on my own technological issues to supporting those of others to teaching classes using technology to managing a large technological organization. Most recently, I've moved on to designing large systems to stimulate innovation and learning. Parallel to this, as a social scientist, I've been exposed to various forms of Systems Thinking, starting with Thomas Kuhn in grad school. As I have pushed the envelope technologically, I have increasingly rubbed up against the systemic constraints imposed by linear, analog thinking. This was a primary impetus for writing this book.

At its root, however, my fascination with systems starts from a very practical standpoint: I want to help people push Digital Age concepts forward in the face of a vast network of Industrial Age systems. Games, being systems for building systems, are one way I have discovered that allows me to help others gain perspective on the paradigms that shape their lives.

The Digital Age has given me the technical tools to create games. These tools have also shown themselves to be remarkably adaptable to constructing environments for the people I work with to work through systemic change. The modern approach to Systems Thinking goes back at least as far as the 1960s with the work of Jay Forrester. (Forrester 1971) However, Howard Rheingold points out that Charles Babbage, a conceptual father of computing, also, ironically, was an early proponent of what we now understand as systems thinking.

> He also devised a means of analyzing entire industries, a method for studying complex systems that became the foundation of the field of *operational research* a hundred years later. When he applied his new method of analysis to a study of the printing trade, his publishers were so offended that they refused to accept any more of his books. (Rheingold 2000, p. 26)

The last point is yet another illustration of why the systemic change we have been discussing in this book is so difficult. One approach to dealing with such intractable problems is by applying principles of play in order to explore ways of advancing change without generating a nullifying reaction. The struggle with analysis and consulting is that you are often trying to teach people who are resistant to learning. Most of us are, especially when it comes from others. Ultimately, we have to learn for ourselves for that learning to leave a deep impression. Reflective play can be critical for engaging participants in self-learning.

Confronting the powerful insights that systems thinking offers us for engaging with complex problems is conceptually challenging, especially when we are a part of those systems ourselves. Marrying play with systems thinking requires an accessible schema to operationalize. In 1999, Donella Meadows published "Leverage Points: Places to Intervene in a System" that seems to give us just such a schema. (Meadows 1999) Meadows proposed that all systemic change is a product of the modification of one of 12 different leverage points. The higher up you go, the more impactful those changes will be on the system.

However, this has consequences. As Meadows's colleague, Peter Senge, wrote in 1990, "the harder you push, the harder the system pushes back." (Senge 1990, p. 58). In Meadows's leverage points model, the closer to #12 you go, the more disruptive you are to the system, the stronger the resistance to said change will occur. Her twelve points illustrate this resistance curve. The lowest level (measurement) engenders the least resistance, while her highest level (surfing paradigms) is likely to be beyond the ability of most systems to absorb, as it requires a systemic questioning of the system itself. Getting audiences to understand the implications of working your way up Meadows's ladder of consequences requires them to step outside of their existences safely. Games can form a mechanism for achieving this because they are an abstraction of reality. The rungs of Meadows's ladder are:

1) Constants, parameters, numbers (such as subsidies, taxes, standards) — *making minute changes to the system by changing the way we measure it*
2) The sizes of buffers and other stabilizing stocks relative to their flows — *how much give the system has to absorb changes*
3) The structure of material stocks and flows and nodes of intersection — *how easily change flows through the system and/or how easily it reacts to stimuli*
4) The lengths of delays, relative to the rate of system changes — *the time that the system requires for registering feedback to action or stimuli*
5) The strength of negative feedback loops, relative to the impact they are trying to correct against — *how strong the reaction to violations of the system are*
6) The gain around driving positive feedback loops — *how sensitive the system is to creating self-reinforcing loops (both good and bad) that will ultimately disrupt the system*
7) The structure of information flows — *does information about change percolate within the system effectively and does it pivot when circumstances change?*

8) The rules of the system — *how the playing field is defined by established rules and who sets those rules*
9) The power to add, change, evolve, or self-organize system structure — *how effective self-organizing ecosystems are within the overall system in changing the overall structure of the system*
10) The goals of the system — *what are the macro-level goals of the system, how are they defined, and who defines them (requires deep realistic assessment: just because they're unstated doesn't mean the goals don't exist)*
11) The mindset or paradigm out of which the system arises — *the basic cultural assumptions of the system (see also Thomas Kuhn)*
12) The power to transcend paradigms — *the recognition that even paradigms are fluid constructs; "if no paradigm is right, you can choose whatever one will help to achieve your purpose"* (Adapted from Meadows 1997)

Meadows clearly recognized that change is not a product of technology or external factors, but that change is a human cognition issue, making it a perfect fit for open-ended gaming. Meadows seems to provide us a fairly approachable roadmap for navigating systemic change. The problem is that even Meadows's schema is not quite simple enough. These 12 points mask some deep considerations that have to be explored in order to achieve structural change. As I set about trying to explain her ideas to my colleagues, my first instinct was to develop a graphical representation of her ideas using the visual tools we explored in Chapter 2.1.

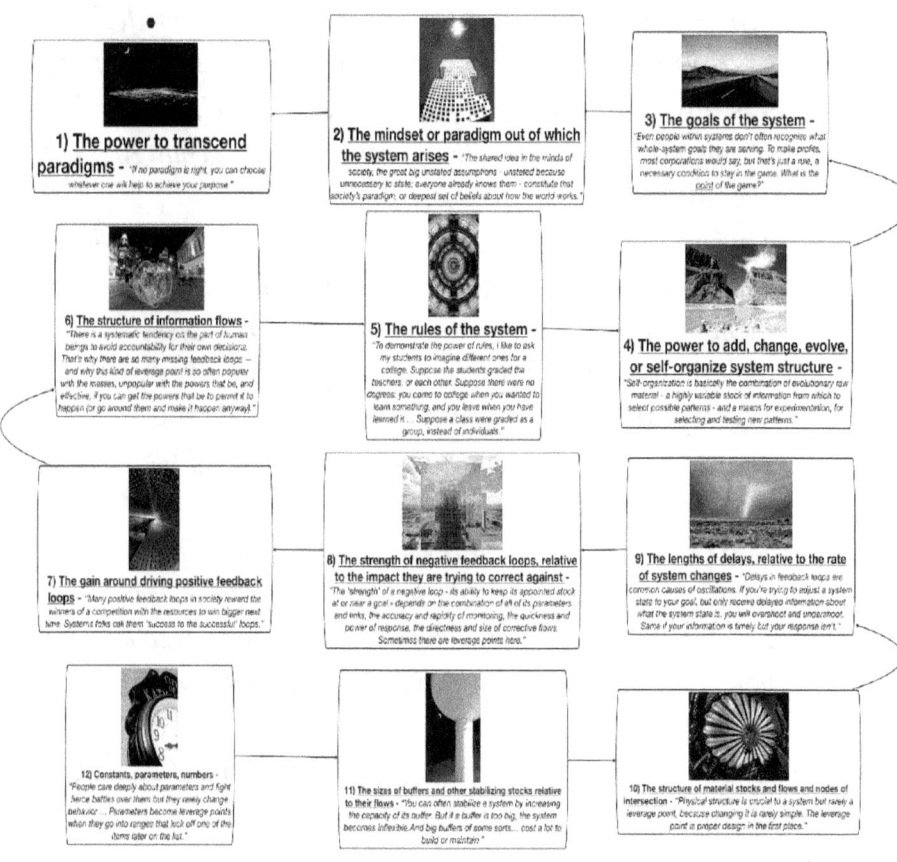

Meadows's 12 Leverage Points as a Concept Map

However, what immediately leapt out to me was this concept map's resemblance to a game board. What emerged was my Paradigm Shift brainstorming game, which allows teams of players to game out any project using Meadows's Leverage Points as challenges to be met and overcome. While it may seem to be a bounded game in how you progress from leverage point to leverage point, how you advance is through open-ended ideation engaging with ever more difficult conceptual challenges in the manner of Chen's DDAs.

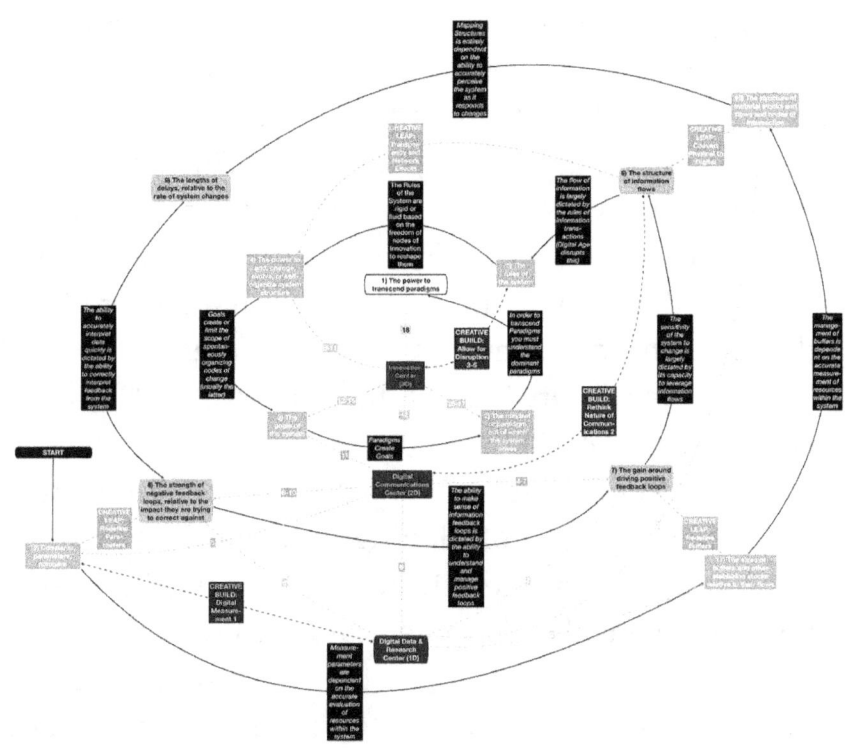

Meadows's 12 Leverage Points as a Game Board

This exercise is an excellent illustration of how exploration of play by leveraging digital tools can lead to the exploration of complex ideas. The game itself can become an engine for structural change. Note how my awareness of digital tools facilitated my progression from text to graphic to gaming scenario in fairly rapid order. We can draw a direct line from the spirit of hacking that we discussed in the first section of this book to the idea of iterative play. My digital tools lowered the costs in both money and time for my experimentation, but they also created a mechanism where I could help others play with ideas. The costs of playing have declined dramatically since Shannon and Thorp. My game development cost me the price of software (about $20) and the cost of some printing and blank cards (also about $20). The design and ideation costs were not insubstantial, but they also served as a mechanism for my exploration of the ideas through the act of game creation. Digital tools reduced the amount of what J. C. R. Licklider would call "clerical feasibility" tasks to maybe 10-20% of my effort. (Licklider 2003, 1960)

This brings us to some of the deeper implications for our exploration of playing with how we go about our daily tasks. It often strikes me how hard it is to get people to consider the potential for play in rethinking how they work, despite the almost trivial cost involved these days. Our industrial mindsets teach us that the costs of failure are high and therefore to be avoided. After Samuel Johnson was writing in the late 18th Century, the dominant thinking in the Industrial Age moved toward a harder and harder separation of play from "work." The industrial mindset did not see play as serious. Unnecessary experimentation also had the potential to be quite costly.

Shannon had the luxury of not caring about the costs of his experiments. The Digital Age has dramatically reduced the costs of these failures to those who do not share Shannon's advantages. The roulette experiment, which cost thousands of dollars in 1960, could easily be replicated with off-the-shelf sensors, a Raspberry Pi, and simulation technology for less than a hundred dollars today. Yet, we continue to shrink back from the perceived "costs" of failure. Why is that?

As Land and Jarman argue, school is good at training us to avoid failure. (Land and Jarman 1993) This is a direct product of the industrial processes that are inherent in spitting out the widgets we call "graduates." Schools are judged by the number of students who achieve fixed, "correct" goals because systems perceive them as minimal requirements for success.

This kind of thinking is a logical result of the perception that work in an industrial context is linear, often hard, and difficult to pivot. Like many features of the Industrial Age, it is also profoundly dehumanizing. Humans learn by experiment and failure. Industrial Age education's high levels of disciplinary specialization retards exploration and new paradigmatic thinking.

Play does not recognize barriers but can be stopped by them. The false consequences created by high stakes testing and even grading undermine all sense of play in a classroom as no one is interested in "playing with their grades," least of all the learners who we need to retrain in the art of play. (Haymes, 2020/2)

The constant grind of industrial metrics, from performance reviews to grades, is profoundly dehumanizing and antithetical to a spirit of play. People feel like they're on a gerbil wheel most of the time, with entirely predictable hurdles and outcomes.

A limitation of the industrial model in both education and commerce is that it is hard to create fluid environments. One example was the shopping mall, which started out as a novel vision and ended up a stagnant soul-killing set of suburban boxes with the same sets of stores and experiences. As Steven Johnson says in *Wonderland*, "[T]he overly programmed nature of the mall environment ended up being its fatal flaw. As always, ***play is driven by surprise and novelty….***" (Johnson 2016, p. 61, emphasis added) Success drives retail marketing. Failure is an anathema to the mall experience. Success in luring people into buying the latest products and experiences is the entire reason for the mall's existence.

Without meaningful failure, there is no ability to play. It's boring to play a game that you know you're always going to win or lose. The investment and stakes in shopping malls are massive. Failure is very costly. The digital environment allows us to create low-cost experiments while maintaining a sense of novelty and wonder. That is why so many of us are addicted to video games. They do precisely that.

The low overhead of digital environments also means that they can easily transcend the kinds of artificial barriers created during the Industrial Age. A digital mall doesn't have to be a physical or metaphorical row of stores. A digital playground can be anywhere and involve anything.

Kids don't argue that popsicle sticks are only food delivery devices. They are also raw materials for arts and crafts projects. Like children, hackers like

Shannon, Engelbart, and Kay did not observe disciplinary boundaries as they pursued larger goals such as beating roulette, making computers into accessible tools, or augmenting human intellect.

Without the interdisciplinary hacking of people like the aforementioned trio, there would be no iPhone, no personal computer, and no internet. They operated outside the usual boundaries of educational establishments. Indeed, all three had non-traditional career paths. None of them fit neatly within their disciplines as traditionally defined. This is because they were out of alignment with educational "disciplines" as defined in the Industrial Age.

The struggle to align educational outcomes with technological realities is a direct product of the specialization that has characterized our systems of thinking for a century or more. The defining characteristic of Industrial Age academia has been the increasing specialization of disciplines. At the highest levels of research, this is necessary, but for the vast majority of undergraduates, these silos are counterproductive. Play disregards silos. Thorp and Shannon ignored disciplinary boundaries between physics, math, and computer science when creating their roulette computer. To get to an iPhone-level device, you have to integrate coding with hardware engineering and design (and ultimately marketing and supply chain operations). You need a multidisciplinary approach to practically any problem these days.

Yet this is precisely the kind of play that Industrial Age education struggles with. Take coding, for instance. Is coding a technical subject taught in a Career and Technical Education environment? Is it an exercise in mathematics and logic best taught by engineers? Most controversially, should we teach coding as a foreign language? The answer is that it encompasses all three and yet most curricula insist on trying to insert it into traditional boxes that don't really create effective coders.

Good coders have to understand logic and the basic technical limitations of the various languages but also need to be able to speak to the machines naturally. Technology companies seek out these kinds of people. Our systems do a very poor job of producing them. And coding is only the first level of the problem. To this point, the best coders were mostly created outside the educational environment because they could play with other subjects. They still find wonderment in their activities because they don't look at them as technical activities but as a means to their creative ends. Their technical skills are often an afterthought, just like conjugation is an afterthought when you master navigating a French railway station on your way to Mont St. Michel or the beach at Nice.

How do we work toward a system that creates these kinds of thinkers? One way to do this is to create environments where it is okay to play. In the 1950s, early computer scientists at MIT and elsewhere often came from the model railroading community. Model railroads are complex electrical switching systems, so the transition was a natural one. Bundles of wires often had to be cut, or hacked, and reconfigured in order to make a system work. These skills translated well into the computing environment of the time. This is where we get the term "hacker." This kind of mindset disrupts established paradigms because it grows out of a general affinity for experimental play.

Over the last 20 years, hacking has returned to its physical roots through technology spaces that allow users to build, break, and repurpose technologies. We have added advanced tools, such as 3D printers and laser cutters, to allow the rapid fabrication of prototypes. Microcomputers such as Raspberry Pi's and Arduinos provide rapid access to electronic brains. Hackerspaces, and their more commonly referred-to cousin Makerspaces, have made their way from community spaces into education and beyond. It is not always a natural fit.

Makerspaces go against the grain of the industrialized educational model. Like coding, they raise questions about where they belong. Viewed purely as a set of tools, they can augment traditional programs in Career and Technical Education or the Fine Arts. In higher education, they are most commonly installed to support engineering or other STEM-based programs. While these programs can benefit from access to these technologies, it is only through broad access to these kinds of spaces that we can cultivate the next generation of Da Vincis. This problem extends to the business world, where spaces must have a logical function and many organizations have trouble wrapping their heads around open-ended outcomes. The employee "recreation" room has become rather popular of late, but it has almost never associated with spaces like a Makerspace, which should be designed for creative play. It's also considered a luxury rather than a central element of the business. Businesses connect the rec room almost entirely to the question of employee retention rather than employee optimization.

Divorcing play and relaxation from work is a fundamental error. This is not an excuse to turn play into work, but the opposite. The most important element here is giving people the ability to play in the manner that Thorp and Shannon could. If playgrounds like Makerspaces are tucked away supporting specialized programs and/or are invisible or restricted to only a few, they are as useful as a library of unread books. If we constrain them by defining "acceptable" activities that can take place within them, they are no longer environments for playing.

The struggle to figure out the applicability of Makerspaces is a clear illustration of the difficulty inherent in grafting a Digital Age environment onto Industrial Age organizations. As we implement spaces for play and creativity, the very nature of the industrial education and business systems work against our purposes. Even after we get rid of resource limitations, structural limitations persist. Again, we return to the central tenet of the book: The problem is not the technology. It is us.

The play that takes place in a Makerspace is inconsistent with the rhythms of the school or business in which it is located because Industrial Age notions of ⌛ **TIME** condition those systems. If you think of education in terms of physical spaces, then you marry yourself to schedules, and the physical logistics of moving people around. This leads to semesters, credit hours, transcripts, testing, and grades. In the business world, this leads to the farce of quantitative performance reviews, which everyone realizes have little or no applicability to the contributions a person makes to profits, which are notoriously hard to trace themselves to a specific individual. These are misguided shortcuts to measure human potential using industrial metrics.

Structures emphasizing play are therefore always going to be in tension with Industrial Age notions of the value of ⌛ **TIME** and ✪ **SPACE**. Neither learning nor innovation is a direct product of "butts in seats." However, we find over and over that these ⌛ **TIME** and ✪ **SPACE** considerations drive almost everything in industrial educational and business systems. They are a product of thinking of technologies of learning and productivity as being scarce and also being non-fungible.

For instance, the classroom is a potential arena for play, but the logistics of industrialized learning severely constrain its potential. First, the need to accumulate class ⌛ **TIME** is a metric used for funding, completion agendas, and other counter-productive avenues for teaching and learning. Tying students and teachers to a fixed place for a predetermined module of learning during a set period of ⌛ **TIME** is an anachronism. We can say the same of the traditional office ✪ **SPACE**. Learning and innovation may happen, but they are incidental. Inspiration and innovation do not happen on cue. These paradigms should be open to question after we saw the impact on learning and work that happened because of the pandemic of 2020. (Haymes, 2020/2)

Teachers have long realized that most learning happens (or doesn't happen) outside the classroom. Learning is therefore only really effective in a sustained environment where students can play with the concepts introduced by teachers rather than exiting the learning environment.

This is why graduate school labs are such fertile creation engines. They are places where people gather to play with concepts, physical objects, and each other. Iconic innovation environments such as Xerox's PARC, Stanford's SRI, and Google are direct outgrowths from this kind of environment, and often emphasize play as part of their work processes. Yet, we constantly create environments in education that are antithetical to this kind of activity.

It is no wonder that we then create graduates who cannot fuse play with work. By changing the dynamics of ⌛ **TIME**, the Digital Age offers us many pathways out of this paradigm but until we have the courage to break out of our siloed mindsets, our systems will continue to be hopelessly out of step with the modern needs of business for playful and creative employees. By disconnecting ⌛ **TIME** from ✧ **SPACE**, digital technologies allow us to more efficiently tune systems around the humans who inhabit them.

There is an incredible amount of wasted time and productivity lost in tying people to specific places and specific purposes. Those constraints limit our ability to play. Coming up with more fluid visions of school and the workplace will create opportunities for Flow and play to happen on a wider scale, leading to deeper learning and more productive thinking. The Digital Age gives us a whole new suite of tools that can enable this kind of holistic approach to our daily tasks.

This dichotomy is yet another product of humans migrating to the factory and becoming part of the machine. Play was deeply woven into the agricultural world as evidenced by the work songs, community work events (see Amish barn raisings), and festivals that were part of the agricultural rhythm of life. These fused play with the hard work necessary to maintain an agricultural existence in a pre-automation era. However, this kind of life faded throughout the Industrial Age as more and more fled the hard life of the country for the perceived opportunities of the cities. This migration came at the expense of play.

It should come as no surprise that play suffers under the industrial mindset. In the Industrial Age, man became part of the machine. The machine did not brook play. As Stuart Brown argues, "In the early days of the twentieth century, industries didn't want workers who could think. They wanted people who could be relied on to repeat the same assembly-line motions efficiently." (Brown 2010, p. 199)

Humans had to adapt to the rhythms of the factory, not the other way around. We measured uptime in machine cycles, not human ones. The most important metric was "more." Only a small elite concerned themselves with "better" and a common perception was that the masses didn't deserve to play like Edison, Bell, and the university elite did. Play disrupted the industrial machine and was therefore to be avoided. Geniuses who did not get the lucky breaks, Nikola Tesla being a notable example, were deprived of the resources that inventors-turned-magnates such as Edison could amass.

Even when machines were minimally involved, such as in the service industry, the industrial mindset was transformative. If you have any doubt, look no further than how the McDonald brothers turned a very non-mechanized effort (the creation of a hamburger and fries) into an automated sequence of events. The McDonald's strategy would find its way into a wide range of service industries and beyond.

If you think about it, the McDonald's mindset has impacted even education as modular degrees are all the rage these days. You do a math module, add pickles, add a writing module, ketchup, etc. and you get a burger, sorry, a graduate. No substitutions allowed. The McDonald brothers and Ray Kroc's innovation was to turn people into machines. Eventually, machines will replace people that can only act as machines. Before that happens, we need to figure out how to retrain them how to play.

The computer, despite its potential as a play enabler, has become one of those machines and our organizations have adapted to that reality because more of the same is what we do with technologies in their first phase of existence. We have realized vast efficiencies and put lots of human cogs out of work in the last 50 years because of computerization, but the vast majority of the remaining workers stayed tied to machine cycles in much the same way as their parents were tied to the industrial assembly line. We now measure their work in bits instead of car parts but, from a mindset perspective, that's essentially the same thing. What we are witnessing in both our work and our schools is the last gasp of the machine age. Workers and students are widgets, measured by their excellence in being widgets, and chucked out of the system just like a burger-making machine chucks out Big Macs, and disposes of the imperfect batches.

The speed of change now, however, is increasingly not rewarding these kinds of outputs. Disruptions are occurring in areas where industrial modes of thinking can no longer keep up with the speed of communication flows. Play is no longer a luxury to be reserved for elite innovators. It is increasingly a necessity required of everyone. It is essential that we learn to leverage Digital Age tools to play with ideas before they overwhelm us. This is true of both individuals and the organizational systems that they shape.

The way we get to an environment that nurtures and rewards play is a profoundly human challenge, not a technological one. In this area, we actually have to shed technologically driven mindsets, those of the Industrial Age, and to rediscover the three-year-old's ability to play. The only way to do this is to demolish false barriers preventing us from playing in our lives, whether those are disciplinary or departmental. There is no longer a logistical excuse to support that (if there ever was one). This will require that we rethink a wide range of our institutions to better (re-)align them with the realities of technology and, more fundamentally, humanity.

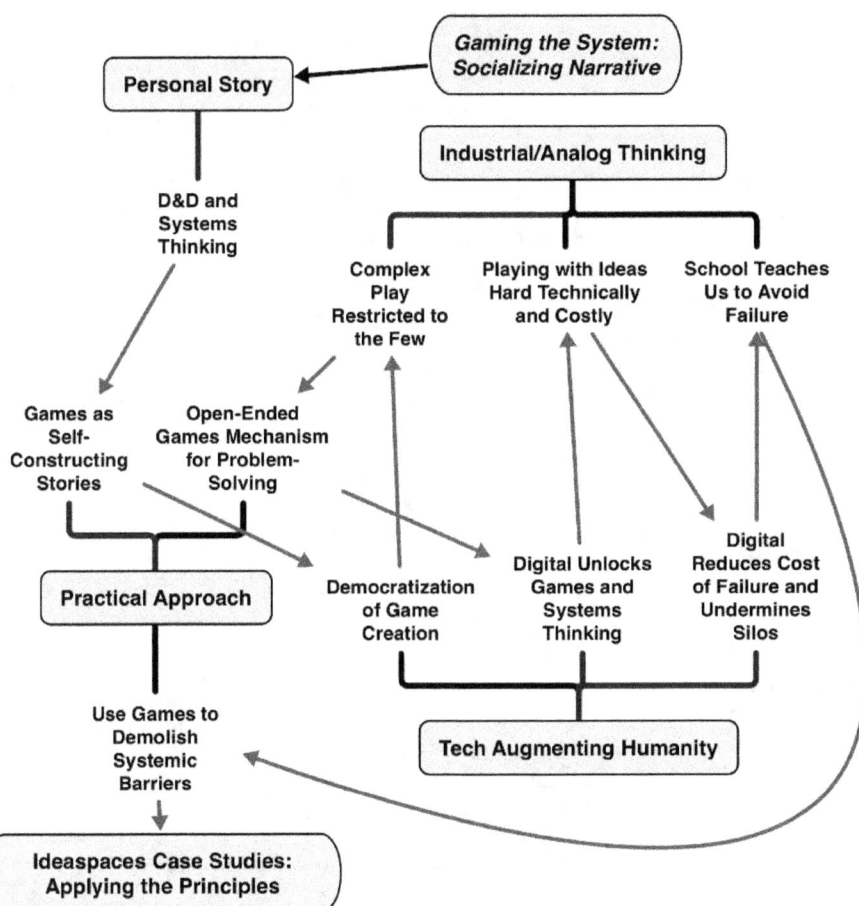

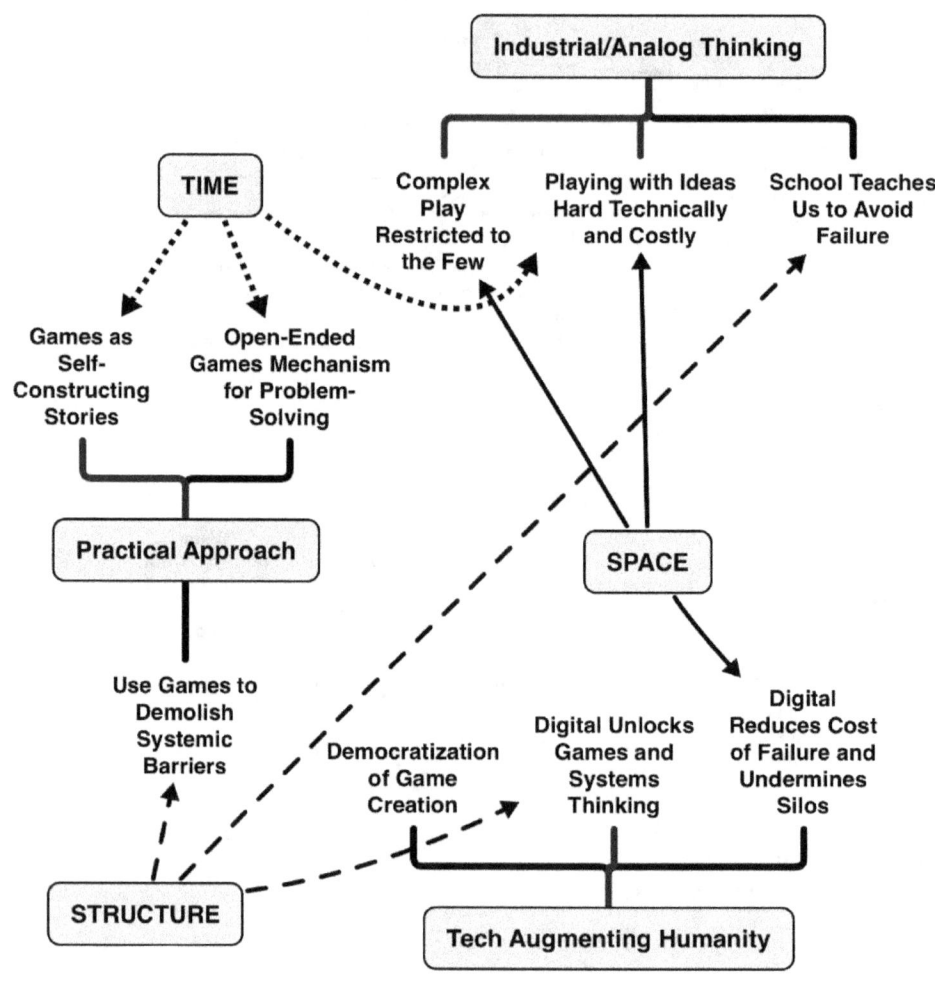

3.0 Designing Human-Centered Technology Systems

When you step into an intersection of fields, disciplines, or cultures, you can combine existing concepts into a large number of extraordinary ideas.

Frans Johansson

3.0.1 Introduction to the Case Studies

When one looks at innovation in nature and in culture, environments that build walls around good ideas tend to be less innovative in the long run than more open-ended environments.

Steven Johnson

The intent of this book is to keep our explorations grounded in reality. We must grasp the opportunities being created by our shift from analog to digital culture. The preceding chapters outlined a key design philosophy that was often missing in the Industrial Age: that you can create human-centered designs that flex and grow with the needs of the humans that will inhabit or use them and that we need to prioritize augmenting our capacities individually and as a species. The needs of *implementing* a technology should never outweigh the ultimate *use* of that technology. This seems obvious in principle, but can be hard to execute in practice.

The IdeaSpaces framework generates a series of questions about any project and how it fits in with the creative needs of the individual and the larger group. How can we create human-centered technological ✪ SPACES? How do we give humans the ⧖ TIME to grow and adapt to change? How do we ✺ STRUCTURE our systems and organizations to facilitate instead of retard growth and change?

These case studies illustrate thinking about challenges through a digital lens. This can have very concrete outcomes in the design of such things as learning and collaboration spaces, or more abstract outcomes such as the design of networks of ideas. All of them have one goal: augmenting the creativity and innovation of the ultimate users, and, by extension, humanity. It is at this intersection where we will find our digital humanity.

3.1 The Antifragile Class: Applying the IdeaSpaces Framework to Redesigning Instruction in a College Government Course

It's the questions we can't answer that teach us the most. They teach us how to think. If you give a man an answer, all he gains is a little fact. But give him a question and he'll look for his own answers.

Patrick Rothfuss

Case Study #1: Integration

👁 **DESIGN PURPOSE:** *The Digital Age requires a fundamental redesign of instruction. Many elements of how we go about teaching are direct products of an Industrial Age paradigm that treats students as widgets. Starting in 2018, I used the IdeaSpaces framework to redesign instruction in my college-level government classes to make them more individualized and meaningful to students who struggle to understand the purpose of many college-level activities.*

✣ **SPACE:** *Enlisting a spectrum of in-person and online spaces, I could shift the modality of the class based on the needs of the student and content, not the limitations of particular kinds of spaces.*

⧗ TIME: *Using a blend of in-person and online spaces, I could redefine the temporal modalities of the class, liberating it from exigencies of scheduling and reducing the load on physical facilities. The major temporal distinction in activities shifted from when everyone could be in the same place at the same time to when was it beneficial for instruction to engage in asynchronous activities.*

❊ STRUCTURE: *There are still many structural elements, such as grades and class sections, that impede the learning experience. However, exercising control over the structural elements of the class by turning it into a semester-long design challenge, I pushed many of these limitations into the background.*

Every semester, I am confronted with a new set of challenges. Some never change. I teach at a large, urban community college. My students are overwhelmingly disillusioned by their experiences in our systems of education. They lack many basic learning skills that more advanced students possess. As a result, they look at school as a chore, a burden, or a necessity and not as an opportunity to explore higher level thinking and concepts. Compounding this, I teach classes that almost no one wants to take. Legislative requirements mandate all students take US and Texas government courses to graduate. On the one hand, this guarantees me a captive audience. On the other, captives are generally uninterested in doing anything more than what is necessary to get them through the experience.

From Day 1 of my teaching experience I was confronted with the challenge of teaching rather abstract concepts—how to construct a democracy—to groups of students who failed to find much relevance in the subject, lacked many of the skills necessary to master the more abstract analytical and critical thinking skills necessary to contextualize the material, and were only there because they had to be. These realities framed the basis of my design challenge. IdeaSpaces provides a lens for unpacking this complex learning environment.

Over the years, I tried many techniques, both technical and pedagogical, to change this equation. However, as long as I maintained the "usual" flow of the class (lecture, discussion, review, test, rinse, repeat), the ❃ **STRUCTURE** itself became the narrative of instruction. Students rarely looked beyond the next assessment (if I was lucky) and quickly forgot the material once the test was over. Efforts to get them to engage in higher-level thinking were always fraught with a lack of foundational work and effort on their part. Whatever richness that occurred in our classroom discussions was not reflected in the work produced by the vast majority of the students. Without motivation or desire, making the effort to learn advanced thinking skills fell beside the wayside, to be replaced by an exercise in triviality.

This outcome never satisfied me. Every semester I made changes I hoped would provide me with a knife to cut through the Gordian Knot of indifference and motivate my students to engage in something that was essential to their roles as citizens in a democracy, if for no other reason. James Madison wrote in 1820 that, "A popular Government, without popular information, or the means of acquiring it, is but a prologue to a farce or a tragedy; or perhaps, both. Knowledge will forever govern ignorance; and a people who mean to be their own governors must arm themselves with the power which knowledge gives." (Madison, 1820) This statement has long been a driving force in why I teach what I do. My students represent the future of our democracy and everything I do as a teacher is geared toward making them successful and active citizens within the democratic ❋ **STRUCTURES** that Madison and his colleagues constructed in 1787.

Intellectual growth aside, if I wasn't engaging my students, I was failing in the mission Madison had set before me. Besides being shorted by the educational system, my students have one other thing in common: they pretty universally come from politically disenfranchised groups that have had little impact on the politics of our country. They, more than most, need the tools that I'm trying to give them.

After returning from a hiatus from teaching in 2018, I determined I would completely redesign my classes from the ground up. Instead of trying to get the students to rise to my level, I would redirect the class to meet them where they were. This meant understanding what was meaningful to each student in life and building connections to government from there.

Providing the physical and virtual ✡ SPACE necessary to nurture self-discovery

✡ **SPACE** provides environments for all forms of communication. As discussed in earlier chapters, we often limit our conception of ✡ **SPACE** to physical environments, but interactions can take place across any set of tools that form ✡ **SPACES** where human conversations can take place. The real distinction between various kinds of spaces is not their physical nature, but *how* they facilitate interactions among the humans who use them. This applies to learning spaces in the traditional sense (classrooms) as well as a wide variety of non-traditional spaces, both online and physically on campus. Informal spaces are at least as important as the limited "in-person" ⧖ **TIME** afforded by meeting for brief intervals in a classroom ✡ **SPACE**. (Haymes, 2020/1)

Digital affordances vastly expand the universe of tools available for teachers to interact with students and for students to interact with one another. Instead of evaluating tools for what they are, I evaluated them for what they could do in contributing to the goals of my courses. In this scenario, a classroom becomes a meeting room, the Learning Management System becomes an asynchronous message board, and assessments take the form of instructor-student communications on progress toward a mutually agreed-upon individual goal.

We often think of ✡ **SPACE** as a series of boxes when in the digital world it is really a continuum. In-person spaces have their own sets of challenges, ranging across furniture, lighting, color and technological limitations. I have spent years trying to create the "ideal" classroom or informal learning area and, in some instances, I think I've come fairly close. However, physical ✺ SPACE is costly and bounded by the laws of physics, geography, and money. Taking my class beyond these considerations involved thinking about digital options to augment what I could achieve in the physical ✺ SPACE. This is not some mysterious process. Any teacher who has assigned readings from a book or homework of any kind has extended instruction beyond the physical ✺ SPACE where the class meets.

The Digital Age gives us a much more sophisticated set of tools with which to carry out this work. For instance, in my class I immediately hit upon the idea of working in public so I replaced the private ✺ SPACE of papers and tests with the public ✺ SPACE of blogs and web pages where my students could see and interact with each other's work. By doing this, I shifted the ✺ SPACE from a ✺ SPACE that only contained the student and myself to one that emphasized a community effort.

The Final Portfolio, the capstone project for the class, is a public website on a platform chosen by the student. This artifact transcended the ⧗ TIME usually bounded by the ✹ STRUCTURAL framework of the course as it was something that they could keep permanently to demonstrate skills learned in the class. It was therefore their ✺ SPACE, not mine. Giving ownership of ✪ **SPACE** is a mechanism for empowering students, which is both a pedagogical strategy as well as a meta-goal for the class itself. ✪ **SPACE** therefore became a tool for creating learning.

When the Pandemic of 2020 hit, it was a fairly simple matter to shift spaces around to deal with the challenges of remote teaching. All of my students' work was already taking place online. The chief challenge was shifting the conversational ✪ **SPACE** of the classroom into another modality. I did this in two ways. First, I took a hard look at content that I was covering in class and determined that I could cover most of this material asynchronously through a combination of announcements, instructions, readings, and recorded videos.

Doing these kinds of activities digitally augmented the tools of persistence and repeatability. Doing this in a live ✪ **SPACE** meant that it was subject to mistakes and omission on my part as I repeated myself from one course section to the next. Having those course elements frozen by technology was actually a good thing.

The second, tougher problem was how to replace the live, interactive community that my classes had evolved into through the narrative space of online communications. Obviously, a videoconferencing platform could provide the ✪ **SPACE** (subject to technical limitations on my students' part). However, assuming that I could simply replicate the conversations that happened spontaneously in a physical space could be replicated around a virtual "table" was a mistake.

Students hide online for a variety of reasons. I did not think it was appropriate to require them to turn on their cameras or even their microphones. On a campus, students enter public spaces. We expect them to conform to certain norms of decorum in that environment. With videoconferencing, however, we were inviting ourselves into their homes and a different set of rules applied. I could not yank them out of their homes without robbing them of agency, a primary mechanism of the class. Therefore, I opened up all avenues of communication in the ✣ **SPACE**. Students could turn on their cameras, microphones, or simply participate through the chat. Most of them chose one of the latter two options.

This additional distance, however, came with its own set of challenges. Socratic teaching was almost impossible as students were often non-responsive, hiding behind their avatars. Instead, I came up with a set of active learning activities that forced the students to engage mentally with what we were working on. I did this by adding additional online spaces, such as live concept mapping, to our synchronous engagements. Getting them to enter the virtual ✣ **SPACE** mentally became the focus of these activities.

Reflecting back through my design process, I can say that my classes are actually stronger than they were before the Pandemic and will probably stay so. This was because I analyzed the spectrum of information and activities with which I was trying to communicate; placed them appropriately within an ecosystem that encompasses a wide range of ✿ **SPACES**; and strategically employed them to maximize their effect. The IdeaSpaces framework showed the way.

Even when instruction shifts back to more traditional modalities, I will still have the digital ✿ **SPACES** that will now more effectively stretch and focus the always constrained ⧗ **TIME** I have with my students in physical ✿ **SPACES**. Managing this ⧗ **TIME** is, however, also subject to the same kinds of analysis I applied to ✿ **SPACE**.

Balancing synchronous and asynchronous ⧗ TIME to build communities of learning

The Pandemic of 2020 also made me evaluate how I approached ⧗ **TIME** with my students. Learning is about conversations. ✿ **SPACE** provides venues that facilitate or retard those discussions. However, ⧗ **TIME** is perhaps an even more critical factor. Conversations take ⧗ **TIME**. Conversations require ⧗ **TIME** to exchange ideas. Ideas require ⧗ **TIME** to be processed into understanding. Teaching is therefore a constant manipulation of ⧗ **TIME** as we consider what is possible during limited synchronous ⧗ **TIME**, what needs ⧗ **TIME** to percolate with our students, and how some students require more ⧗ **TIME** and repetition to internalize content than others. The Pandemic of 2020 upset these temporal parameters, perhaps even more than it did the spatial factor of proximity. Fortunately, I had already begun considering precisely these factors long before remote instruction struck.

One of the first things I did with my class when I set out to redesign it in 2018 was to deconstruct how conversations happen. There are four different speaker-audience nodes of conversation: one-to-many, one-to-one, many-to-many, and many-to-one. It is a mistake to think that these spontaneously occur even in a classroom where all the participants are in the same ✪ **SPACE** at the same ⧗ **TIME**. Any teacher that is honest with him or herself recognizes that the vast majority of conversation in most classes is one-to-many.

It is also a mistake to believe that most learning happens within the walls of the classroom even though most students like to operate as if this is the case. That is because there is a temporal modality to add to the equation. Some learning happens synchronously. Some learning happens asynchronously. None of these factors are new. We've always assigned homework, readings, and projects to be completed outside of the physical meeting of class. However, applied creatively, the Digital Age allows us to play with all of these different modes.

This is not a new thing, but few teachers that I have observed over the years took advantage of the repertoire of tools at their disposal to shift ⧗ **TIME creatively**. This is difficult even for me. There was constant frustration with students not doing their work outside of class (which caused me to feel like I had to cover the content as lectures) instead of a systematic examination of what the most appropriate use of class ⌛ **TIME** was. This caused a deformity of ⌛ **TIME** and created a focus on the limited meeting times students and teachers were together in a physical ✺ **SPACE**.

This perception was at the root of much of the instructional panic when the pandemic forced human separation. Remote learning unmasked how ineffective synchronous content delivery was. Teaching Socratically proved to be particularly ineffective when dealing with groups of students online. Individual students turned out to be quite resistant to communicating through a video conferencing platform.

Digital technology makes everything more transparent. Classes were recorded and students' unwillingness to engage in instruction was made more visible than they might otherwise have been in traditional classes. The teaching that was usually confined to the walls of the classroom could now be seen by anyone who wandered by their children sitting in front of the computer or watched the recording later.

Another factor that put pressure on our learning systems was the sudden loss of informal spaces and the lack of foresight among those developing Learning Management Systems (LMS) in somehow replicating the hallway and library within the LMS. A lot of learning takes place in informal spaces, especially when coupled with on-demand synchronous support services.

Fortunately, before the pandemic struck, I had already been working for several years on structures and techniques within my in-person classes designed specifically to engender conversation rather than preaching. I could break down the functioning of conversations in my class and create the opportunity, at least, for spontaneous learning on the student's terms. I even embedded a librarian in my class to offer support for informal discovery.

This process started with a deconstruction of the activities within my course, recognizing the limits of structures and technological environments. I did this while still meeting with my classes in a physical setting. While this required some shifts when instruction went remote, I was already halfway there because of my ongoing IdeaSpaces analysis. Breaking down my class by audience and conversational modes (synchronous vs. asynchronous), I could adjust strategies accordingly. Here is how I deconstructed the conversations that take place in my class to get me there:

- **One-to-Many** - This consists mainly of class announcements and content. These benefit from asynchronous access and repeatability, so recorded video or documents are the best media here. The pandemic forced me to take the ⧖ **TIME** to create much of this content. Having it done will improve my classes in the future because I can concentrate on teaching, not preaching.

- **Many-to-Many** - My students work in public, so many-to-many is a central part of their workflow. Using discussion boards as a way of collecting and sharing resources is one example of this. Blogging and creating virtual products subject to peer feedback is another. Again, these activities benefit from persistence, so most of it is created and reviewed asynchronously. Periodic group synchronous review also shows the commonality of students' struggles with the material.

- **Many-to-One** - Review and feedback are essential to any model based on formative assessment. However, this is hardest to replicate remotely. Students have to be prodded to offer constructive critiques of their peers. This benefits from a congenial class/group environment. It is difficult to establish that level of genuine rapport if no one is ever in the room together.

- **One-to-One** - This is where small class sizes become critical. The success of massive online efforts like the UK's Open University can in part be attributed to the small tutorials that form its distributed core. (see Weinbren, 2017 pp. 132-133) I was lucky in that the modality I was teaching in during the 2020/21 academic year was limited to 18 students per section. I could therefore meet with each student individually at least once and, in most cases, twice over the course of the semester. This allowed me to give them individualized feedback and allowed me to hear from them their challenges, no matter where they were, and adjust learning strategies to help them overcome whatever those were. The removal of most ✪ **SPACE** and ⧗ **TIME** considerations through low overhead videoconferencing allowed me to meet the students where they were. I merely had students sign up on an online spreadsheet to a ⧗ **TIME** and scheduled meetings throughout a week, where I canceled regular group meetings.

As I describe more extensively in *Learn at Your Own Risk*, I could bend ✪ **SPACE** and ⧗ **TIME** to teach during the pandemic. There is a lot of overlap between the two. However, recognizing the affordances of both allowed me to pivot and adjust to a variety of circumstances and needs.

Understanding the differences between synchronous and asynchronous ⧗ **TIME** spent in the class, plus the different transmitters (authors/speakers) and receivers (audiences), is critical in structuring instruction, no matter the circumstances. The Digital Age offers a vast panoply of resources for changing the constraints on communication. Breaking down these activities within a course of instruction was critical to bending ⧗ **TIME** to my advantage as a teacher.

However, ✸ **STRUCTURE** still gets in the way. Administrative decisions about class size and sections are difficult to overcome. I had to face this challenge from the perspective, not of a program manager, but from that of a teacher trying to exist within frameworks and paradigms that students, faculty (disciplines), and administrators create. That limited how much I could bend things in my design process.

Counteracting ✸ STRUCTURES that get in the way of effective learning.

From my position as a teacher, it is difficult to construct the systems under which students are required to operate. The best I can do is to create systems within my class that push back on some of the more pernicious effects of grades, compartmentalizing courses, and artificial times for learning. These are the frameworks that guide students throughout their educational careers and are very hard to overcome. Pressure from above to observe certain standards in order to collect "metrics" that supposedly measure student achievement are a constant source of interference.

✸ **STRUCTURE** is about giving communication meaning. Humans construct paradigms to make sense of the multitude of conflicting stories around them. Learning and education are no exception to this. Well-designed structures allow us to build knowledge and skills systematically. The challenge is creating structures that contextualize learning into a broader educational and life journey for the learner. School should be about acquiring meaning, not credentials. Industrial educational structures have their own systemic logic, which often has little to do with constructing knowledge. All too often, systemic pressures devolve "education" into a dehumanizing bean counting exercise. Digital Age technologies can help us transcend that effect. This was always an explicit structural goal of the redesign of my course.

The first step in this process was to recognize the structural constraints under which conversations with my students' labor. Perhaps the most destructive of these is the tyranny of grades. Industrial Age education systems generate the perception that, like a job, education is a transactional process: You give me your time and money and I will give you credit and credentialing in the form of a degree. Layered on top of this are incentives to rank order students based on a standard categorization of competence.

My class therefore devolves into a sophisticated form of bribery where I exchange grades for jumping through my structural hoops. I call this "transactional teaching" and cover it extensively in a chapter in *Learn at Your Own Risk*. (Haymes, 2020/2, pp. 59-70) Transactional teaching is extremely pernicious when trying to generate real meaning in my students' learning. We have trained students that the grade represents the product, not intellectual growth. Therefore, activities that are not structurally incentivized through some sort of reward system do not get done.

Developing systems to counter this is a tricky proposition. On the one hand, paying people to go to an intellectual gym will still result in growth. However, many psychological studies have shown that, once we remove that reward, the incentivized are far less likely to continue that activity. That is why I have never given my students academic credit for voting, for instance.

But how does one get around this in a ❋ **STRUCTURE** so warped by broken incentive systems? My patch is to reward all activity within the class while constantly reminding them how these activities contribute to the larger capstone project. I have also eliminated arbitrary hoops, such as tests that merely exist to get students to perform for the teacher.

My only summative assessment of student work (one where I rank order the quality of student work as the basis for my grade) is the Final Portfolio. All of my grades before that point are incentives for doing a specific level of activity. I try to push subjective assessments to the side as much as possible. The student does the level of work required to meet a minimum threshold and they get the credit for doing it. I also use an additive point ❋ **STRUCTURE** rather than a deficit-thinking averaging ❋ **STRUCTURE**. The idea here is to shift away from deficit thinking toward additive thinking in measuring their achievement. They can look at accumulated points and compare them to the remaining points available in the semester. It's not about what they failed to do, but what they still need to accumulate.

The only summative assessment in the entire class, the Final Portfolio, comprises a website the student owns, and which persists long after the class is over. I emphasize to them that this is something they own, a tangible product of their learning. It is also something they can show to employers and others as a demonstration of their competency in argumentation, analysis, technical presentation, and communication, all highly sought-after skills of any college graduate.

Digital Age tools make this much more possible through their persistence. I can show students how the strands of their learning fit together. They can keep a legacy of their work far more meaningful than a letter on a transcript.

One victim of the pandemic has been our connection to reality. Many students have complained that it is precisely this kind of connection that they were missing the most because of remote instruction. Building something tangible has proven to be something my students can hang onto even when everything else seems to be unraveling. They seem to respond. Almost half of them earned A's and B's in the Fall 2020 semester and my drop and failure rate was at or below my pre-pandemic levels. I was able to structure the activities so that the class responded in an antifragile manner to crisis. There is no reason to doubt that these same strengths will persist and flourish when things return to what passes for "normal."

There are other systemic constraints such as the siloing of disciplines because of ✿ **SPACE**, and ⧗ **TIME** that Digital Age tools can help us overcome. Critically, systems also have to be set up to facilitate informal learning, as I described earlier in this chapter and discuss in more detail under the *STAC Model*. (Haymes, 2020/1) Playing with all kinds of ❋ **STRUCTURES** becomes far easier when we lower physical restrictions on ⧗ **TIME** and ✿ **SPACE**. Just as I brought a librarian into my online class, I could bring in outside experts or work with other professors to merge competencies into true interdisciplinary efforts. Digital Age liberation from spatial and temporal constraints can therefore form a strategy to undermine anachronistic structures. We can then create structural *opportunities* for learning. Digital ❋ **STRUCTURES** offer our best hope for creating the kinds of citizens and thinkers that will be necessary to meet the much larger challenges beyond the pandemic, such as automation and climate change.

CONCLUSION

In the example of my role as a teacher in the ❋ **STRUCTURE** of my class, I could build upward from ✿ **SPACE** and ⧗ **TIME** considerations to influence structural outcomes. It would be a lot easier if those structures supported my efforts to pivot my students' learning to more closely align with the realities of the Digital Age. However, this case study shows we can employ Digital Age tools to instigate guerrilla-like structural changes by modeling alternative systems and highlighting anachronisms within our Industrial Age educational structures like grades and disciplines. The next case study illustrates the attempted construction of a Digital Age ✱ **STRUCTURE**, the West Houston Institute, and the challenges and shortcomings of engaging in such a high-profile systemic challenge.

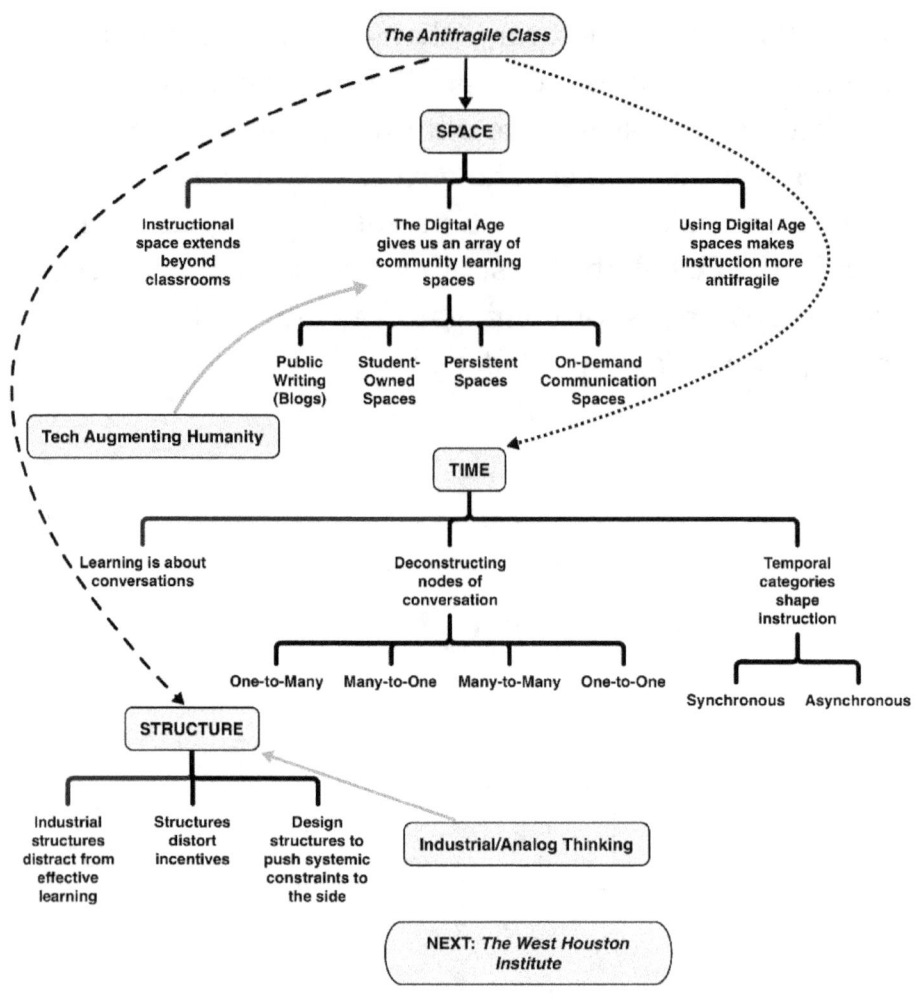

3.2 The West Houston Institute: Design for Augmentation

No attempt has been made to achieve the character of a university campus with its separate buildings... On the contrary, all buildings have been connected so as to avoid fixed geographical delineation between departments and to encourage free interchange and close contact among them.

Oliver Buckley, Bell Labs, 1938

Case Study #2: Application

👁 **DESIGN PURPOSE:** *The West Houston Institute catalyzed change efforts at a community college with 75,000 students and over 5,000 faculty and staff. The team built it with design principles in mind that made the building itself and the programs it supports subject to evolutionary (emergent) design practice.*

✹ **SPACE:** *The West Houston Institute creates seamless transitions between technology and human spaces. These spaces form integrated environments that support human augmentation activities at scale.*

⧖ **TIME:** *The WHI plan envisions a series of integrated programs leveraging technology to enhance their impact and reach.*

❋ **STRUCTURE:** *There are explicit modalities designed to spread the impact of activities taking place within the walls of the WHI to the larger Houston Community College system.*

The West Houston Institute (WHI) is a $50 million innovation building that was heavily influenced by the IdeaSpaces concepts. We designed it as a tool to help people transition to digital thinking and adopt design principles as outlined in the preceding chapters. The frameworks of the WHI are sets of tools that augment the learning and innovation missions of the campus. The ultimate design of the building sought to implement three intertwingled goals for its inhabitants as part of a holistic concept encompassing a recognition of the opportunities and challenges created through a Digital Age assessment of ✪ **SPACE**, ⧖ **TIME**, and ❋ **STRUCTURE**. These goals have succeeded or failed based on the level at which they challenged existing systems.

1. Leverage the principles and tools of Making, both physical and digital, throughout the building
2. Create intersections in an environment specifically designed to break down barriers among the various college constituencies and with the larger community

3. Reinvent teaching and learning for the Digital Age by combining innovative classroom and informal ✿ **SPACE** design into a holistic learning environment

In 2018, Houston Community College (HCC) opened the West Houston Institute, an innovative campus designed to facilitate design in all of its forms. This 100,000 square foot building incorporates a Makerspace, a Digital Media Center, experiential science labs, a teaching innovation lab, a conference center, and a wide range of collaborative spaces. We designed these spaces to grow mindfully and integrate holistically. We also explicitly designed the WHI to facilitate the human aspects of the process of design. The ✿ **SPACE** creates the kinds of human collisions necessary for innovative design to happen.

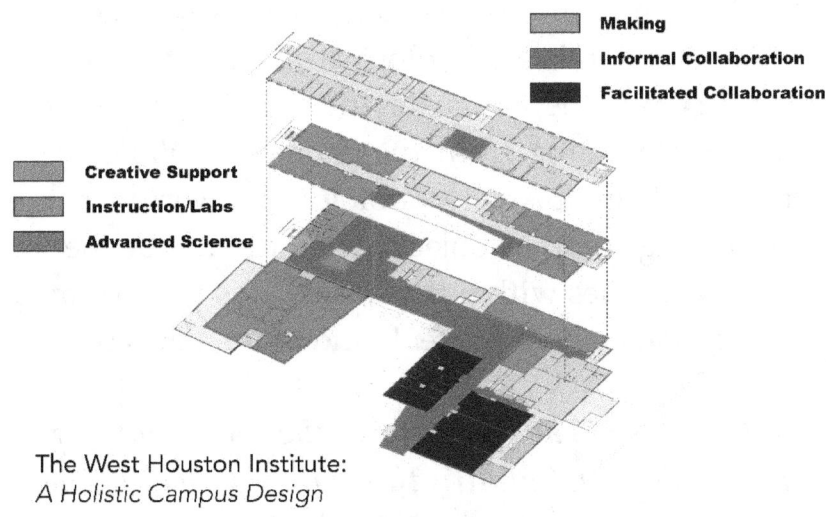

The West Houston Institute:
A Holistic Campus Design

A functional plan of the West Houston Institute

The building has at its center the concept of Making. It is through the act of creation that students learn that life doesn't come with an instruction manual. Often in the process of Making the creator(s) will discover that the pathway to a successful design will involve multiple disciplines and approaches to an ever-changing set of challenges. While any portion of the building can become an area where Making can take place, the WHI has two areas specifically tasked with different aspects of the making process: the Idea Studio (Makerspace) and the Digital Media Center.

The first of these is the 10,000 square foot Idea Studio Makerspace itself. The design team toured a range of Makerspaces that encompassed a broad range of design goals. We wanted to fold many of these together to develop a design that would support a wide range of creators that included (but were not limited to) artists, engineers, entrepreneurs, and the local community of Makers. The key element of Making is not the technology itself. The technological ✪ SPACE is the arena for the creation of ideas and objects stemming from those objects. While we can leverage Digital Age technology to lower the barriers to creating complex objects, it should not be the focus of the activities within the spaces. It is in the minds of the humans where the real innovation takes place.

✪ SPACE: *The spaces form the tools that augment the inhabitants' ability to create and learn. They do not in themselves create this effect.*

We designed the Makerspace to be open to the rest of the building through extensive use of glass. The intent here was to create interest from the casual passerby and to lower the exclusivity and mental barriers to entering the ✿ **SPACE**. We found through its prototype Makerspace, the Design Lab, that students and other members of the learning community seeing what was going on in the ✿ **SPACE** led to curiosity and a desire to get involved in the unique activities taking place there. This is the true purpose of the ✿ **SPACE**, not the "cool" toys that might be located there. This is also in keeping with the overall design philosophy of bringing together a diverse set of people and interests to create a "Medici Effect" and spur innovation throughout the Institute.

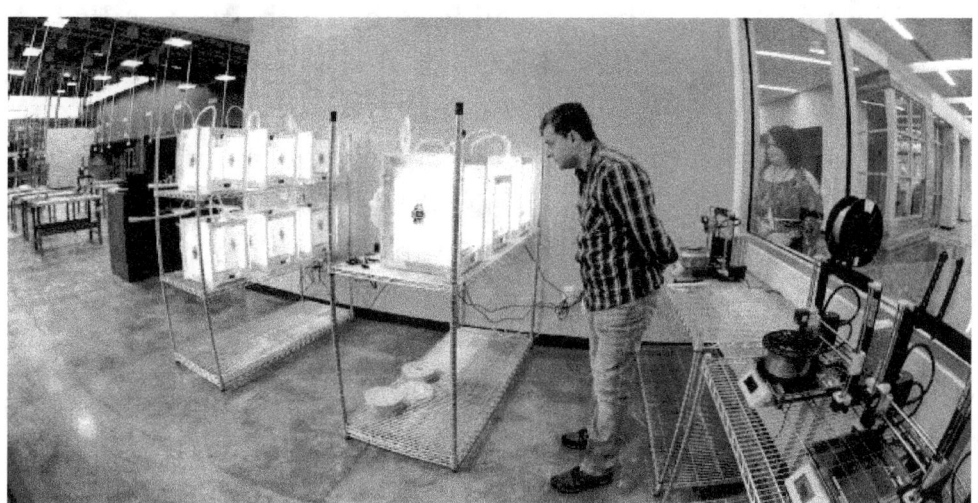

WHI Makerspace

Immediately adjacent to the Makerspace is the Digital Media Center, which supports the creation of digital artifacts in the way that the Makerspace supports the creation of physical artifacts. The Digital Media Center contains a One-Button Studio, an audio recording booth, a digital projects computer lab and staff who can help both faculty and students to create multimedia content. Through experience and earlier piloting of spaces, HCC discovered that this digital content was implemented in a wide range of activities, including pitching entrepreneurial projects birthed in the Idea Studio. However, it is also a low-barrier entry into the creation process.

Digital tools are relatively easy to access and manipulate. Mistakes carry very low consequences. Messing up a poster, video, or website results primarily in a loss of time (and that is ⧗ **TIME** well-spent learning how to recover from failures), whereas messing up a project in the Idea Studio might cause the loss of physical materials. While the consequences of that are low, they can still intimidate in a way that a purely digital project might not.

✧ **SPACE:** *The explicit connection between virtual and physical creation is an intentional part of the design of the two spaces, which are also physically next to one another.* ⧗ **TIME:** *Programmatic activities within these two spaces are often explicitly linked as a level of familiarity with computerized design tools is a precondition of transferring those ideas to tools that will produce a physical or virtual "product."*

At the other end of the building are advanced experiential science labs and classrooms designed to create an explicit connection between science and creative pursuits. Experimentation using the Scientific Method forms a core to all effective design and entrepreneurial projects. Like the Makerspace areas, the science labs at the WHI are designed to be experienced by a broad range of students, not just those specifically studying science and engineering. The advanced science lab is being set up to carry out actual undergraduate research through the eventual installation of a Scanning Electron Microscope, an X-Ray Diffraction Device, and an Atomic Force Microscope. We will port activities in this lab out to video displays in the hallway. Science students will also be encouraged to collaborate with activities taking place in the Makerspace and elsewhere throughout the campus.

Collaborative areas connect all the spaces in the building. At the largest end is a conference center that is the building's connection to the outside world. It hosts events that showcase the work of students and faculty created within the Institute. The conference center brings in outside thinkers who inform and inspire the community of innovation and learning that we see emerging from the WHI.

Next to the conference center is the Collaboratorium, a facilitated collaboration area. Between these spaces and the Makerspace/Digital Media Center is a range of informal collaborative spaces designed to give students many venues for informal interactivity. The design of these spaces influenced the development of the STAC Model. (Haymes, 2020/1)

☼ **SPACE:** ***It is important to recognize how adjacent spaces can influence the impact of the tools in an area. Having informal collaboration spaces next to the Idea Studio, for instance, allows students to bring work outside the*** ☼ **SPACE** ***and create new, serendipitous connections with other activities in the building.***

What will emerge from the WHI is unknown. Unlike other design projects, it has no end. It will always be a work in progress. We designed all spaces within the building for maximum adaptability as new roles and needs emerge. Stewart Brand quotes Brian Eno in his book *How Buildings Learn,* "An important aspect of design is the degree to which the object involves you in its own completion." (Brand 1994, p. 11) This quote sums up the mission of the WHI on many levels.

The Digital Age creates opportunities for organic and evolutionary growth. It also opens up intersections that were previously hard to discern. In this quote, a technologist (Brand) is taking a sentiment from a musician (Eno) and applying it to the concept of architecture. The Digital Age makes such connections easier to discover. At its root, when working as designed, the WHI is a participatory building that directly involves its inhabitants in its evolution.

By incorporating design as an explicit part of the WHI ❁ STRUCTURE, we have made it somewhat more resistant to systemic backlash.

There were, however, very real design considerations that led to this conclusion and outcome. The team recognized early on that many of the technologies that were being designed into the building were at risk of being obsolete before we even opened the doors. Therefore, we made the explicit decision to design the building to evolve. We built classrooms with extra space to allow for them to adapt as teaching methods changed or even to be converted into project spaces to support efforts in the Makerspace and elsewhere. The connecting "arteries" of the building were built extra wide to allow for changes in how they might be used. There are very few spaces that were not designed with this level of flexibility in mind.

This was not as easy a case to make as it may sound on its surface. It goes against many of the fundamental budget-driven practices that have emerged around building design and construction. Since cost control is the dominant paradigm in this world, budgets and equipment allocation are based on very specific use case scenarios. These may (or may not) be updated with new technologies, but the practice is the same. You need this 3D printer. It takes this kind of power and floor space and therefore the wall and power grid need to be built around those constraints. If that is outside the budget, you don't get the 3D printer, but you also don't get the infrastructure built around the now-theoretical printer and lose future capability as requirements change.

This is an analog, Industrial Age approach to the problem. It assumes a fixed set of solutions, not a flexible set of outcomes. It was critical in our vision of continuous adaptation that the Makerspace and other technology spaces were designed to capacity rather than around a specific technology set. The ceiling in the Makerspace is completely open to the deck. We can easily move power and network around to support new devices when and where they are needed. It is possible to cut additional ventilation into the roof structure should that become necessary.

However, as our technology needs shifted even during the design process, there was always a systemic urge to erase the underlying infrastructure when the design team eliminated a particular device. For instance, we deleted an industrial 3D printer at one point that required 220-volt power in favor of more, smaller printers that ran on standard 110-volt power. The engineering team wanted to delete the 220-volt capacity in that space. I had to step up and argue that, while the solution we had pivoted to assumed a lesser infrastructure (which was already in place), that they should not delete the more robust infrastructure because it was entirely plausible that we would need it at a later date and time. The Digital Age, because of its rapid shifts in technology, demands that you build to a level of capacity and let the technology elements fall into place around them and be replaced as needed by the benefits of technological innovations and shifts.

✡ **SPACE:** *Design decisions need to be made holistically and should always take advantage of unexpected opportunities, knowing that rapid shifts of technology are not only possible but almost certain to occur.* ✻ **STRUCTURE:** *Design processes should operate as holistically as possible. Focusing on small parts of the design can lead to shortsighted decision making.*

This building supports humans and human creativity. At the heart of this idea is an emergent design process that addressed a wide range of activities. Design is a collaborative activity. As Tim Brown, CEO of IDEO defines design, it is, "a human-centered approach to innovation that draws from the designer's toolkit to integrate the needs of people, the possibilities of technology, and the requirements for business success." (Brown) Much of the ✡ **SPACE** within the WHI should serve as a toolkit for collaboration, whether that be in collaborative forms of teaching and learning in the classrooms and beyond or the acts of collaboration necessary to create new ideas. The Idea Studio is a toolset designed to bring people together. So is the Digital Media Center. The building itself should be explicitly seen as a tool in the facilitation of human interactivity.

✪ **SPACE:** *Tools facilitate fundamental human activity.*

Human interactivity can take place in a wide range of settings and within a wide range of structures. The nature of the various collaboration areas within the WHI reflects a wide diversity of needs required to bring people together in order to promote change and growth. The Makerspace is as much a collaboration ✪ **SPACE** as it is a technology ✪ **SPACE**. Making is a collaborative process, focused on people, not their tools. At the other end of the spectrum, there are spaces where the technology recedes even more into the background. The Collaboratorium is a facilitated brainstorming ✪ **SPACE** and complements a host of informal spaces in between.

The building facilitates an intentionality of collaboration across a broad range of people and groups. As Frans Johansson writes in *The Medici Effect*, breakthroughs are, "a result of different people from different fields coming together to find a place for their ideas to meet, collide, and build on each other." (Johannsson 2004, p. 12-13) This insight explicitly informed our design for the Institute. We wanted to create a range of spaces and supporting technologies where diverse groups of people could come together and interact in both formal and informal venues.

*Informal Collaboration in the
West Houston Institute's Collaboratorium
(Photo: Laura Williamson)*

In order to promote the interaction of people and their ideas, we looked at Google's attempts to maintain an innovation culture through spaces explicitly designed to facilitate creative collisions. Google has discovered that knowledge can only really spill over if there is a surface – like a chalkboard – on which it can land. If two people run into each other in a hallway and spark a conversation that requires technical discussion, or idea generation, or any topic that might benefit from being written, it is best if they can write it down immediately. Since hallways are not conducive to stopping to write, Google offices are strewn with "every conceivable gathering space, from large open spaces to tiny nooks with whimsical furniture." (Molloy 2013) Conversations never take place too far from a ✿ **SPACE** to gather.

One of the early design stipulations we made was that there would be no hallways in the building. We held to this idea on the first floor and also allowed for significant collaboration areas on the second and third floors. The areas in dark gray on the floor plan illustrate all the informal collaboration spaces. They range from small, even individual, areas at the top left of the floor plan to the large conference ✿ **SPACE** at the bottom right. The spaces in between include a wide range of furniture that can be deployed as pods, collaboration tables that include televisions and computer hookups, as well as computing resources scattered throughout the common areas. There are also plans for mobile whiteboards to be scattered throughout the areas to create spontaneous work surfaces for the students and the WHI community to use on the fly. Additional informal collaboration spaces can be found on the upper floors to allow similar spontaneous collaboration to occur.

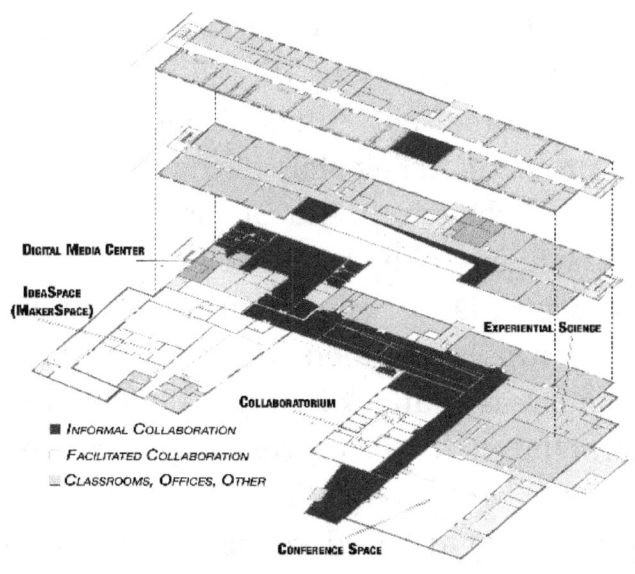

Informal Collaboration at the WHI

We typically think of Makerspaces as technology spaces. This is an inaccurate view of what makes them work properly. They are actually collaboration spaces deploying advanced technologies. These are strategically combined to facilitate the realization of ideas into physical objects, such as prototypes. At the WHI, the Idea Studio's large assembly space is an open work environment where projects can be built individually or collectively. The concept is for ideas and solutions to spill over throughout the ✺ **SPACE** and for entrepreneurs, artists, engineers, and others to come together in developing solutions for real projects. All the while, they are learning from each other.

The Collaboratorium is a facilitated brainstorming environment. Groups of up to 32 can gather here for day-long programs using design thinking, creative problem solving, and other techniques to attack wicked problems. They start in the large Convergent ✪ **SPACE** at the south end of the area. The smaller spaces facilitate the breaking up of groups to deconstruct the issue being discussed into smaller bits. Larger groups can then break up into smaller Divergent areas to attack individual parts of the issue. Finally, at the end of the process, they reconvene in the Convergent areas to reassemble the individual solutions. A cloud-based collaboration environment allows groups to work apart and yet have all of their solutions added to a central canvas, which can be accessed and changed (if that is desired) even after the in-person session ends.

The WHI Conference Space (Photo: Laura Williamson)

The Conference Center is a collaborative bridge to the outside world for the WHI community. Outside speakers bring their knowledge to the Institute. Institute participants can share their achievements and insights with the outside world. This ✿ **SPACE** allows the creative communities both inside and outside the Institute itself to interact through a wide range of activities, including MakerFaires, robotics and other STEM competitions, as well as various brainstorming activities. The conference center acts as a conduit for ideas to flow into and out of the West Houston Institute.

❋ **STRUCTURE:** *The conference center and Collaboratorium act as key intersectional conduits to the larger HCC system, demonstrating the utility of the innovative operations taking place within the WHI.*

The Collaboratorium and conference center allow for the WHI to situate itself within the larger communities that surround it, both inside and outside the college itself. They form intersections connecting the intersectional nodes *within* the building to nodes elsewhere. As Steven Johnson said in a 2010 interview, "it's very, very rare to find cases where somebody on their own, working alone, in a moment of sudden clarity has a great breakthrough that changes the world." (Burkeman, 2010) We designed the Institute to give diverse groups of people the opportunity to work and learn together to change the world. Collaboration is a profoundly human activity. It is through our human-centered design philosophy that we have designed a ✿ SPACE that is centered on the humans creating within it.

Even with design principles of the WHI in place, there is still no guarantee that the humans that inhabit the ✿ SPACE will take full advantage of them. I have seen teachers enter the most innovative classrooms and turn all the desks so that they are facing the "front" of the room and proceed to teach in a very traditionalistic manner. The Makerspace is always at risk of becoming a "3D printer training room" and any collaboration ✿ SPACE can be turned into just another "conference room" to the exclusion of any creative activities that might occur there. That is why it is critical to pair these spaces with a set of programmatic activities designed to help humans adapt to the new modalities of the ✿ SPACE. The WHI business plan explicitly laid out two programs, the IDEAS Academy to teach students design thinking techniques and the Teaching Innovation Lab (TIL) to teach faculty how to engage in their own design processes far reinventing how they teach and learn. Both programs assume Digital Age modalities made possible by the facilities of the West Houston Institute.

⧗ **TIME:** *We should orient programmatic activities toward giving humans the opportunity to reflect and adjust to the new opportunities and challenges created by the Digital Age.*

Programmatic activities are more vulnerable to disruption because of funding and systemic constraints. Of the two programs, the IDEAS Academy has come closest to its intended vision, albeit not in its actual shape. In order to test it, we were forced to take it outside of the normal systemic constraints of higher education. We did this by tying the initial cohorts to the adjacent Early College High School of Alief Independent School District. As a senior capstone event, the program taught students design principles. They then designed projects that had a direct impact on their community. However, where the program came up against hard limits is when we attempted to scale to impact college students as well. The reason: it butts up against Industrial Age degree programs and doesn't fit into any of the existing academic pathways for HCC students. The IDEAS Academy is currently a flourishing island separate from most activities within the WHI.

✱ **STRUCTURE:** *The success of programs like the IDEAS Academy leads people to question why it can't be scaled to larger systems. Over time, this will cause them to question the efficacy of existing structures and explore changing them.*

The Teaching Innovation Lab (TIL), however, represented a more direct challenge to existing systems and so has never fully been realized in practice. It was an iterative strategy to incorporate design principles and experiential techniques into reshaping instructional practice. While there had been notable attempts to create a dialog of change at HCC, the TIL represented the first rigorous and comprehensive initiative designed specifically to address the culture of teaching and learning.

The central premise of the TIL is that humans create knowledge. We cannot transmit it. The best we can hope for is the transmission of information. Master teachers lead their students to the creation of their own knowledge within a constructive framework. We designed the TIL to create a framework to allow teachers to become the kinds of learners they hope to create.

To achieve this, the lab created an interdisciplinary environment similar to that of a technology incubator where new concepts could be tested, rapid iteration could occur, and results could be analyzed rigorously. These results were to be collected into a repository of knowledge and experience that could be shared with the larger educational community. Each year, several interdisciplinary cohorts of 12-15 participants would take part in the TIL, getting release time to innovate and collaborate with one another, but still teaching. Everything was to be subject to testing, from pedagogical methods to the spaces and technology that create the learning environments in which we teach. Sharing would be a key part of this process. The TIL was designed to serve as an umbrella for events, workshops, and conferences to support teaching innovation.

⧖ **TIME:** *The lab was specifically designed to address issues of* ⧖ **TIME.** *Its reward system was based on allotments of* ⧖ **TIME,** *not money, and it was created to develop a network of information sharing to scale and sustain its activities.*

Giving teachers the ⧖ **TIME** and ✿ **SPACE** to analyze and question their teaching strategies, and working together to do so, is critical in developing and sustaining a culture of change among the larger community. The TIL sought to create a "liquid network" to institutionalize dynamic and ongoing change. (Johnson 2010, pp. 43-66) A central focus of this community was the development of a recognition of the reality that the Digital Age both provides tools to make reinvention possible and external shifts that make it necessary for the long-term survival of the educational system as we know it today.

We were not unaware of the challenges that were facing us on a systemic level. As Peter Morville astutely pointed out, the challenges facing the creation and functioning of a design environment for systemic change in teaching and learning can be severe.

> Double-loop learning in organizations is rare. Defensiveness in cognition and culture makes it hard to question basic beliefs. Successful people and organizations are the worst, as they've never had to learn from failure. Experts and executives alike deny the problem, shift the blame, and shut down; and the organization runs efficiently off a cliff. We

> can get better, but it takes commitment. We must confront the assumptions behind our ideas. We must surface conflicting opinions and recast them as hypotheses to be tested in public. And must be willing to critique and change our goals, values, frameworks, policies, and strategies. (Morville 2014, p. 96)

For the changes that we envisioned the West Houston Institute to catalyze systemic change, they had to be scalable, cumulative and adaptive. It was also clear that these kinds of changes would meet resistance the more ambitious they became. Therefore, we designed the building itself as an emergent design project. We always recognized that the building would have to adapt and flex as it navigated the white waters of systemic change that were underway. We were, therefore, under no illusions that many of the more ambitious projects we had in mind would be easily accepted or even understood by the larger systems that they impacted. The explicit design of the project was focused on anticipating and adapting to change. I was particularly mindful of this passage from Stewart Brand's *How Buildings Learn* throughout the design and construction process.

> Almost no buildings adapt well. They are *designed* not to adapt; also budgeted and financed not to, constructed not to, administered not to, regulated and taxed not to, even remodeled not to. But all buildings (except monuments) adapt anyway, however poorly, because the usages in and around them are changing constantly. (Brand 1994, p. 2)

We wanted to make a building that adapted well instead of poorly. To the extent we succeeded, we built the building to flex and change in response to changing programmatic needs. All the bones are in place, even if a lot of the muscle is missing. The structural elements are 90% in place.

It is the higher-level programs that address issues of ⌛ **TIME** and ✲ **STRUCTURE** that are the least fleshed out. However, these programs are also consciously designed to be lightweight. They have very few startup costs now that the building is complete. It comes as no surprise that the industrial paradigms that characterize higher education and HCC in particular would not embrace the digital nature of the campus. The campus itself is merely a tool, an element of ✪ **SPACE**. Tools are more often than not ineffective in dictating their usage. If you hand someone a computer and all they see is a nail-driving device or maybe a calculator, that's what they will use it for. The WHI may contain many innovative spaces, but if all they are viewed as is surplus classroom ✪ **SPACE**, that is how they will be used.

✪ **SPACE:** *Even a building is nothing more than a tool and a container for technology and spaces. It cannot create* ⌛ **TIME** *or the* ✲ **STRUCTURES** *necessary for human or organizational change in isolation.*

The West Houston Institute is an excellent illustration of how a system impacts digital transformation. The construction of the tools or spaces is at the low end of Meadows's leverage point scale. Programs fall somewhere in the middle, but the ultimate purpose of the building challenges structural paradigms. It's a lot easier to showcase beautiful or clever design, but that should not be mistaken for genuine change, which can only come when the humans themselves adapt. No one in particular at HCC has "got it out" for the WHI. It's the collective "this is how it's done" attitude that makes progress so difficult and slow.

As long as there is a core group of people at the Institute who understand the design process nature of the building, it can still grow as a catalyst for change. The structural efforts going on within the building represent the kinds of processes that will cause sustainable, systemic shifts at an institution. Those shifts have to occur in peoples' heads, not in technology. The basic tools are all there. Some of the more advanced tools are still being adapted to day-to-day use, but that is also a process of human adaptation and almost never a technological one.

My experience in the business world tells me that the same logic applies there as well. It may feel revolutionary and, from a money perspective, it's fairly easy to change out all the furniture and reorient the walls of the office into much more collaborative designs. These may even result in some incremental changes in productivity if done correctly.

However, long-term structural changes are less clear and are therefore harder to achieve. This is precisely what happened to me at the architectural firm in which I operated for a period. Based on my experience designing the West Houston Institute, they hired me to work on programs to help transition the office and its educational customers to better use digital ⌛ **TIME** and ✧ **SPACE**. I made some headway in some discussions about ✧ **SPACE**. However, after only three months of work on the structural-level changes dictated by the Digital Age, they determined they were not ready for the kinds of questions this initiative raised around their fundamental processes, which were still rooted in Industrial Age thinking. We therefore determined that my efforts in this area would not be productive except at a superficial level. After that, I continued to work on programs geared toward reflection ⌛ **TIME,** but began planning an exit strategy. Nine months later, they eliminated my department.

The architectural firm example is an illustration of how even applying Digital Age tools can raise profound structural issues in much the same way that the West Houston Institute example raises questions about the trajectory of higher education. Some of these issues, such as the inability to scale the IDEAS Academy, have deep structural roots. The next chapter's example, The Deep Thought project, is an example of using Digital Age tools to address structural elements directly.

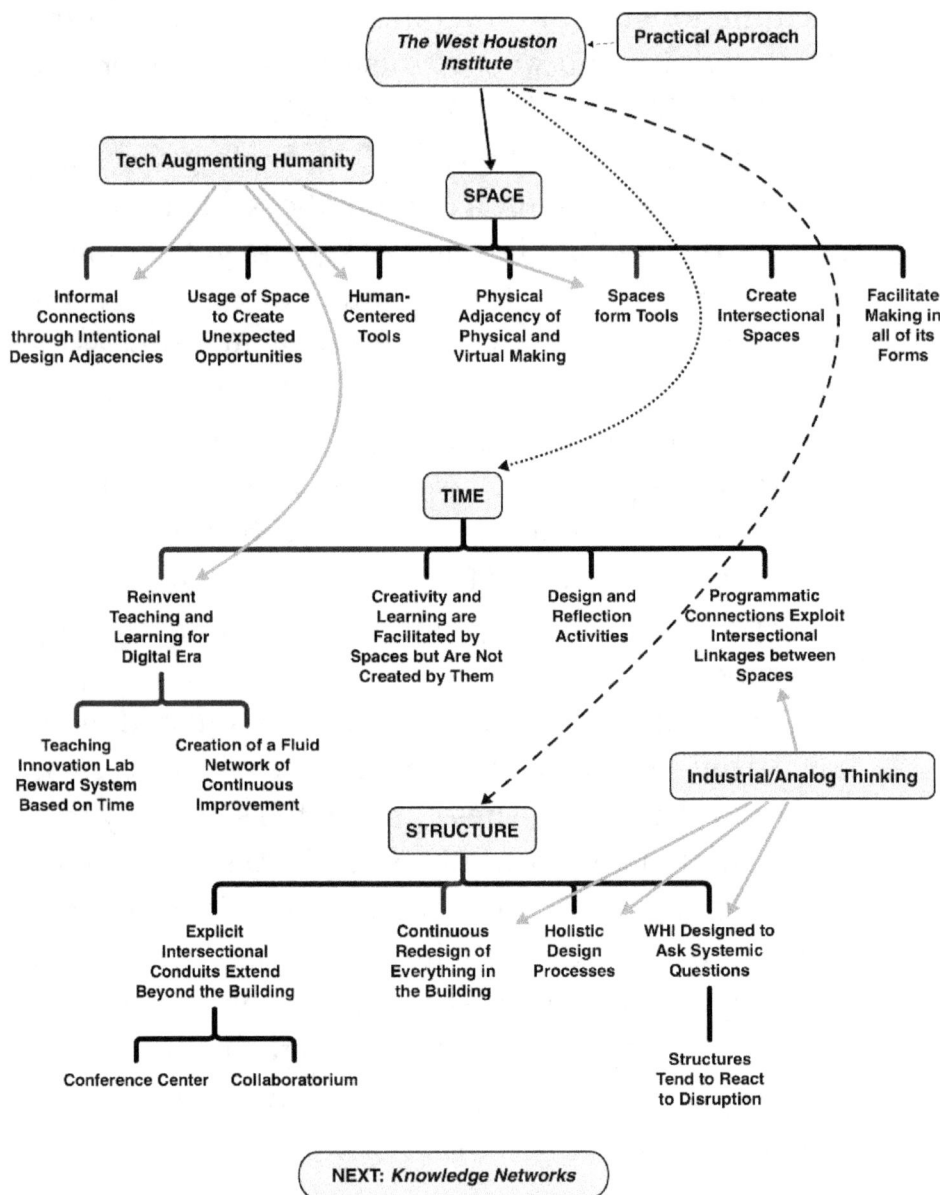

3.3 Thinking Backward: A Knowledge Network for the Next Century[‡]

The world is a system of ever-changing relationships and structures.
Ted Nelson

Case Study #3: Aspiration

👁 **DESIGN PURPOSE:** *Higher education communicates achievement through transcripts, which are based on grades that are inconsistent across institutions, create a perverse incentive structure for students, and perpetrate many aspects of the Industrial Age education model that are inconsistent with the Digital Age. We need to rethink how this knowledge network works in order to stimulate systemic change.*

✹ **SPACE:** *Understanding the implications of digital* ✹ *SPACE is critical to reimagining our knowledge exchange networks.*

⧗ **TIME:** *The Deep Thought Knowledge Network removes analog* ⧗ *TIME barriers from our knowledge communities.*

❄ **STRUCTURE:** *Deep Thought cannot work unless it is aligned with current* ❄ *STRUCTURES of academic discourse and credentialing. It reaches back to older, pre-industrial models of discourse and legitimacy to do so.*

[‡] This chapter was originally published in *Current Issues in Education* 22/1, January 7, 2021. It has been edited for this volume.

Introduction: Stepping Through Paradigms

The world of education creates, shares, and processes information according to established sets of rules. This is necessary or we would have informational chaos. However, the underlying paradigm of technology has shifted even as education attempts to hold on to familiar patterns developed in a world where Newtonian physics was seen as the highest calling. Einstein's Theory of Relativity disrupted this paradigm over a century ago, but our informational paradigms have been much slower to adapt.

The Newtonian worldview was straightforward with rigid sets of rules and structures. This fit well within the Industrial system of thinking that required a high degree of human coordination and conformity. Conversely, the relativistic world is malleable with unexpected bends and gravity holes. This kind of thinking undermines rigid patterns of thought and information exchange because rules are conditional on circumstance. Arguably, this is a closer parallel to how humans, and groups of humans, think. However, our knowledge systems have not kept up with physics.

As early as 1945, just as he completed the Manhattan Project, the great relativistic endeavor of the 20th century, Vannevar Bush recognized we were living in a relativistic information environment but still trying to cope using Newtonian tools of thought. Douglas Engelbart and Ted Nelson, working in subsequent decades, envisioned that the general purpose, networked computer, with its ability to connect and recombine information endlessly, would provide a key bridge over the discontinuity described by Bush. However, more than a century after Einstein, most of our knowledge systems continue to be based on a Newtonian paradigm, even as our supposedly fixed points of information become ever more relativistic.

Einstein's theories took over a decade to be accepted. Information paradigms are even more resistant to change because we cannot experimentally confirm the disruption of the old paradigm like Arthur Eddington observationally confirmed relativity in 1919. We still think of the computer as an electronic version of stacks of paper. This is in part because of the desktop metaphor that Xerox implemented in the 1970s that was extended to the concept of web "pages" by Tim Berners-Lee's version of the World Wide Web promulgated in the 90s. As Nelson lamented, "A document can only exist of what can be printed." (Nelson, 2009, p. 128) Despite the limitations of this metaphor, it is unlikely that the vast majority of computer users who flocked to the new technology in the 80s, 90s, and beyond would have been able to process the metaphor that Ted Nelson had in mind as he was contemplating Xanadu and Thinkertoys in the 1970s.

The paper metaphor was a necessary bridge to introduce computing into the existing knowledge paradigm. In doing so, however, these metaphors sped up the information processing ability of society without augmenting its knowledge processing ability. It is like the idea of a paperless office that was the rage in the early 1990s. It should have surprised no one that precisely the opposite happened when we gave everyone access to a printing press. The computers and networks of the 1990s and 2000s have made us tremendously efficient in handling information, but have not systematically improved our ability to turn information into knowledge.

Like Bush, Ted Nelson and Doug Engelbart were always more focused on the knowledge-creation aspect of computing than they were on its information storage capabilities. They took those for granted. Instead, they were struggling with a new language designed to push us beyond linear, textual Newtonian thinking. In the process they left behind tantalizing tools, from hyperlinks to graphical user interfaces to the concept of the internet itself, for bootstrapping (to use Engelbart's term) ourselves into new ways of thinking about the world. The tools have been eagerly adopted and adapted to achieve efficiencies in the existing linear paradigm, but their true potential has remained elusive.

From the 1500s to the 1800s, humanity underwent a profound transformation from magical thinking to linear scientific thinking to explain the world. Magical thinking is something that every teacher struggles against even today as he or she tries to teach what we now call mental disciplines such as critical thinking and the Scientific Method. As rational modes of thinking are increasingly challenged by the relativistic information environment we find ourselves in, this struggle will become even more intense.

Most of what we are trying to teach in our undergraduate classrooms today can be dated back to nineteenth-century concepts. There is an objective reality, at least in the sciences. Even in areas where objective reality is harder to achieve, such as the social sciences and humanities, there are tools and structures that guide us toward understanding. Until you get to the outer edges of science and math, this remains true. However, the outer edges of that paradigm have resided, at least since Einstein, in a far more ambiguous structure. As Leonard Shlain argues in *Art & Physics*:

> Aristotle, Bacon, Descartes, Locke, Newton, and Kant all based their respective philosophical citadels upon the assumption that regardless of where you, the observer, were positioned, and regardless of how fast you were moving, the world outside you *was not affected by you*. Einstein's formulas changed this notion of "objective" external reality. If space and time were relative, then within this malleable grid the objective world assumed a certain plasticity, too. (Shlain, 1991, p. 136, emphasis in original)

Now, if you imagine a grid of information instead of a grid of space-time, relativistic forces also affect the fixity of rationalistic indexing systems of knowledge. Arguably, this is what we are experiencing today. There is so much information flooding our systems that many conceivable stories can be shaped out of cherry-picking information without context. This is making a mockery out of rationalistic methods and structures of organizing knowledge. Pre-rational theories such as the Flat Earth Theory and others are actually experiencing a resurgence because of the blindness created by a lack of information perspective. Information literacy is a constant challenge in this environment.

We need better technology to lead us out of the swamp lest we sink into it. Rationalism increasingly finds it difficult to be a "candle in the dark," as Carl Sagan so eloquently described it in *The Demon-Haunted World*. (Sagan, 1997) Our brains struggle with rational thought and regress easily into magical thought. As we have done for millennia, we need to develop technological solutions that will allow our species to survive and these need to start with information if we hope to develop answers to the next set of human challenges on the horizon. Our most urgent need now lies in creating a technological solution that will lead us to a paradigmatic shift in how we ❈ **STRUCTURE** knowledge, thought, learning, and inquiry.

Vannevar Bush's Challenge

In prescientific times, people or buildings being struck by lightning were seen as the actions of a vengeful God, even if there was no logic behind those actions. We explain tragedies such as children with autism these days as the actions of vengeful scientists, even when no rationale for such actions exists. We dismiss science for what it doesn't know instead of appreciating it for what it reasonably knows. The unknown and opaque are most easily explained by magic. We easily fall back into magical thinking patterns when we can't seem to find a rational one. Even what science knows is obscured by the vastness of the information that flood its networks. This was already apparent at the height of the rationalist paradigm when Vannevar Bush wrote "As We May Think" in 1945.

Bush's challenge to us in 1945 was directed, not at technology, but at the limitations of human intellect and our societal capacity to manage knowledge. He designed (but never built) a technological solution, the "Memex," with the fundamental purpose of using machines to help us organize and, most importantly, to connect our collective knowledge. He laments that, "Professionally our methods of transmitting and reviewing the results of research are generations old and by now are totally inadequate for their purpose." (Bush 1945, 2003, p. 37.)

In the intervening 75 years, nothing much has changed other than the speed and volume of information we must process. Academia is still largely stuck in an analog paradigm. Information is still, often artificially, siloed. Instead of using digital technology to create systems of knowledge that are associative and re-combinatorial, we have used it to create new and better walls. Instead of organizing knowledge and creating new pathways to solving our ongoing and persistent open-ended problems (to Bush's worry about the Atomic Age, we can add climate change, education, the maintenance of democracy, and dealing with the speed of technological change), we have created a bewildering array of disconnected data that is easily challenged and undermined. Instead of taking advantage of technology to broaden our horizons, we have doubled down on existing cultures of specialization. We have created highways when we should have been creating webs.

Douglas Engelbart, Ted Nelson, and many others iterated on Bush's basic idea of connecting knowledge and used that as the basis for attempting to shape technology in the intervening decades. Yet the interconnectedness that was at the core of the visions that brought us humanized interfaces with our technology and even the internet itself has provided a constant source of disappointment to these thinkers.

This is because true "Networked Improvement Communities," as Engelbart put it late in his career, have always proven to be tantalizingly out of reach. At its core, both thinkers are trying to explore a new language that challenges the text-based linear Newtonian thinking that is deeply ingrained in our academic institutions and beyond. We have been willing to grasp the "shiny toys," from hypertext to graphical user interfaces, that their thinking has led us to. However, we have failed to understand the deeper implications of the paradigmatic shift in thinking that they were trying to show us. We continue to struggle with the misshapen technologies that have descended from that time.

The realities of our information environment are increasingly disconnected from our ability to address them. This is because we are in the midst of a profound shift in the volume and intensity of information available to us. We lack the necessary tools to manage them effectively. Our inability to process these shifts in information volume and intensity is the root cause of many of our problems. We also overlook the depths to which this will shake the foundations of our societies in education and beyond.

It's easy to look back and analyze how previous paradigm shifts, such as the invention of the printing press in 1453, have impacted the shape of the world. However, it is unlikely that someone living in Weimar in 1520 would have perceived the tectonic shifts that this sudden proliferation of pamphlets would wreak upon their society or the terrible bloodshed that would mark the next 150 years. The religious wars of the 16th and 17th centuries, sparked by the paradigmatic shift brought upon Europe by sudden democratization of information away from the Catholic Church, are something that would be globally catastrophic today. Even in 1945, Bush saw the danger.

> The applications of science have built man a well-supplied house, and are teaching him to live healthily therein. They have enabled him to throw masses of people against one another with cruel weapons. They may yet allow him truly to

encompass the great record and to grow in the wisdom of race experience. He may perish in conflict before he learns to wield that record for his true good. Yet, in the application of science to the needs and desires it would seem to be a singularly unfortunate stage at which to terminate the process or to lose hope as to the outcome. (Bush 1945, 2003, p. 47)

As Bush foresaw, information is both the source of the paradigmatic shift before us and our only hope for surviving it. This can easily evolve into a meta-meta discussion about the importance of designing webs of knowledge in order to understand webs of knowledge, but that is precisely where we are today. Despite the warnings of Bush, we are still woefully deficient in our ability to leverage technology to connect diverse strands of knowledge. This results in a fundamental lack of perspective in understanding and addressing the complex problem sets we need to address, whether that is designing educational systems more attuned to societal and economic realities that face our graduates or to create open-ended efforts to address the realities of climate change. We live in a society rich in information and poor in the connective tissue necessary to contextualize it.

Unlike Bush in 1945, we now have the means to design systems to help us see the world in new, relativistic ways. A thinker in 1520 might have perceived the future with the information tools we have today. However, even then, he would not have been able to map an alternative course without the power to contextualize that information. We have at our disposal a set of tools unimaginable to this Renaissance thinker. These tools have the power to provide us with new contexts that change the way we see. As Bush points out, "The abacus, with its beads strung on parallel wires, led the Arabs to positional numeration and the concept of zero many centuries before the rest of the world; and it was a useful tool—so useful that it still exists." (Bush 1945, 2003, p. 42) In other words, the creation of a practical technology led to an unexpected conceptual breakthrough.

Unlike the Renaissance thinker, in conceiving the Memex, Bush mapped out a conceptual framework but could not make it real. Bush lamented that the technologies that he envisioned for the Memex were still tantalizingly out of reach. The potential for creating an abacus-like technology that would lead to unexpected outcomes is much more possible today, however.

We can, if we wish to, finally realize the vision of the Memex and open up unexpected opportunities to see the world in different ways. We have the abacus but are just using it to add numbers together instead of changing numerology. After Bush, Engelbart and Nelson sought to couple systems that would augment thinking to tangible technology in order to address complex problem sets. While they invented radical new ways to apply technology to human problems, they never really undermined the paradigm they sought to subvert.

Subverting Index-Based Thinking

Our thinking systems have never evolved beyond the indexing systems pioneered in libraries and, already in 1945, only seventy years after the first promulgations of the Dewey Decimal System, Bush realized how antithetical these systems were to the way our brains worked. He intuitively recognized that, like many other products of the industrial age, these kinds of indexing systems forced humans to adapt to the machines, in this case card indexing systems, rather than the other way around. They force us into linear thinking processes that are inefficient at best and misleading at worst, as well as being subject to institutional biases. His vision for a Memex addressed this discontinuity.

> Our ineptitude in getting at the record is largely caused by the artificiality of systems of indexing. When data of any sort are placed in storage, they are filed alphabetically or numerically, and information is found (when it is) by tracing it down from subclass to subclass. It can be in only one place, unless duplicates are used; one has to have rules as to which path will locate it, and the rules are cumbersome.... The human mind does not work this way. It operates by association. With one item in its grasp, it snaps instantly to the next that is suggested by the association of thoughts in accordance with some intricate web of trails carried by the cells of the brain. (Bush 1945, 2003, p. 44)

One of the central challenges lies in our technological ability to perceive and reshuffle patterns of information dynamically, a laborious task in a world constrained by paper. It parallels the struggles that science has had to overcome when the dynamic properties of Einstein's relativistic world come into contact with the much more rigid rule structures of Newton.

The challenge that relativity poses to Newton's universe is that it discards much of the "indexing" system that serves as the foundation for mechanistic physics. After Einstein, the basis for physics could no longer be perceived as a fixed fundament. Instead, it became a system of underlying and varying relationships that could have unexpected effects on the structured physical world that Newton laid out in his masterpiece, *Principia*. It doesn't invalidate the notions of Newtonian physics, but it adds an extra layer on top of it.

Likewise, post-rationalism doesn't have to reject rationalism. However, it highlights the importance of looking beyond the fixed points of information and rigid parameters that exist in Newtonian mechanics. In a relativistic world, it's the relationships that matter much more than the nodes that they connect.

Bush perceived this problem and saw in his Memex a tool for solving it. However, the vision presented in "As We May Think" tells us what the system *should* do, but doesn't effectively explain *how* to accomplish it. For instance, how exactly do you share "trails" of knowledge? His system of what amounts to annotation can have tremendous value, but it only goes part of the way in creating new opportunities to view a dynamic grid of knowledge in both ✿ **SPACE** and ⧖ **TIME**.

Seventeen years later, Douglas Engelbart proposed a novel approach to the problem in "Augmenting Human Intellect." In his proposal for "A Research Center for Augmenting Human Intellect," Engelbart imagines an electronic version of the paper-based card indexing system he had constructed to more dynamically manage his information network. With an electronic system, Engelbart envisioned that this could assume a far more dynamic environment in which the information could exist.

> These statements were scattered back through the serial list of statements that you had assembled, and Joe showed you how you could either brighten or underline them to make them stand out to your eye—just by requesting the computer to do this for all direct antecedents of the designated statement. He told you, though, that you soon get so you aren't very much interested in seeing the serial listing of all of the statements, and he made another request of the computer (via the keyset) that eliminated all the prior statements, except the direct antecedents, from the screen. The subject statement went to the bottom of the frame, and the antecedent statements were neatly listed above it. (Engelbart 1962, 2003, p. 105)

By deprecating the "serial" record, Engelbart is pointing out the fundamental inadequacy of text (and paper) in his analysis of how to "augment human intellect." His next act should be seen as a way to struggle out of this box (with 1962 technology, I might add). It is most revealing. In order to explain how his program should work, he gives us this diagram.

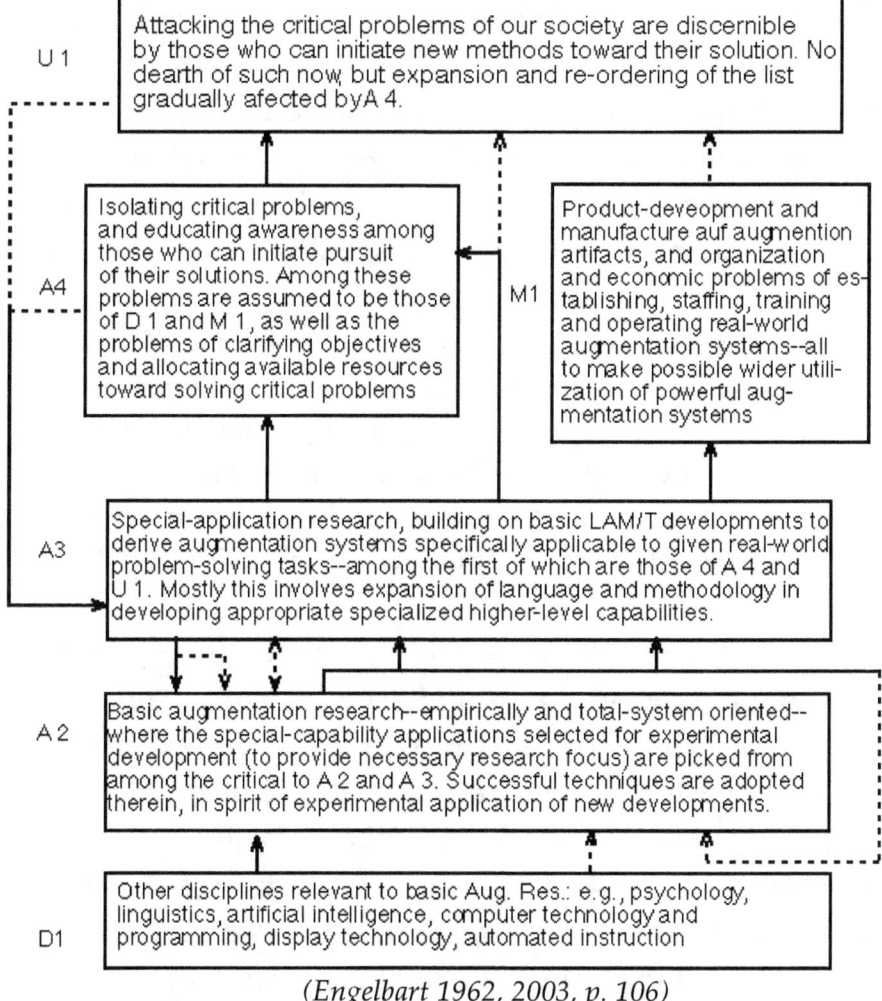

(Engelbart 1962, 2003, p. 106)

Engelbart breaks free of the constraints of text in this diagram in order to illustrate an iterative map forward. "When you get used to using a network representation like this, it really becomes a great help in getting the feel for the way all the different ideas and reasoning fit together—that is, for the conceptual restructuring." (Engelbart 1962, 2003, p. 106) Instead of giving us a fixed textual stream, computers open up the possibility of viewing the relationships in knowledge visually. For the first time in this diagram, we can perceive Bush's trails. Constructing this in 1962 was undoubtedly a manual process, but Engelbart was not one to let technology dictate his larger vision. He clearly saw this as a dynamic possibility when he says,

> "It is a lot like using zones of variable magnification as you scan the structure—higher magnification where you are inspecting detail, lower magnification in the surrounding field so that your feel for the whole structure and where you are in it can stay with you." (Engelbart 1962, 2003, p. 107)

This is not about inventing mice. It is about opening a gateway to thinking on the level that Bush demands. It is about abandoning archaic text-based indexing systems for something more appropriate to a digital vision of the world. It is about creating a visual language as a first step toward creating a relativistic conceptual language. Engelbart thinks that the old way of working with information (text) is severely limited. He saw that technological systems offered a way out of that.

> I found, when I learned to work with the structures and manipulation processes such as we have outlined, that I got rather impatient if I had to go back to dealing with the serial-statement structuring in books and journals, or other ordinary means of communicating with other workers. It is rather like having to project three-dimensional images onto two-dimensional frames and to work with them there instead of in their natural form. Actually, it is much closer to the truth to say that it is like trying to project n-dimensional forms (the concept structures, which we have seen can be related with many many nonintersecting links) onto a one-dimensional form (the serial string of symbols), where the human memory and visualization has to hold and picture the links and relationships. I guess that's a natural feeling, though. One gets impatient any time one is forced into a restricted or primitive mode of operation—except perhaps for recreational purposes. (Engelbart 1962, 2003, pp. 108-09)

Engelbart articulated this vision almost 60 years ago. Yet "modern" information storage and retrieval systems, including the current version of the internet developed by Tim Berners-Lee, have continued to double down on the library indexing model of information sharing. While there are dynamic linkages, the fundament of the internet's addressing ❋ **STRUCTURE** still views sites as a sequence of pages that metaphorically might as well be stacks of paper on millions of desks. Sure, it's nice to have access to all of those desks but, like the warehouse at the end of *Raiders of the Lost Ark*, this plethora of information is only useful if you can connect it up and recombine it in new ways. The site/page metaphor resists atomic manipulation of its contents, sometimes implicitly, sometimes explicitly.

Like this book itself, the metaphors of text and paper lead us down certain pathways in understanding the connections between ideas. One of my struggles as its author has been to pull together a complex web of breadcrumbs into a linear narrative. This skill is important, but at the same time, we have to recognize we must discard all of that nonlinear information for it to be presented in this form. Text forces us down linear channels rather than creating constellations of ideas. Textual tools limit our opportunities to perceive an exploding universe of narrative pathways instead of shifting our larger perceptual paradigms. Without context, information never becomes knowledge.

Ted Nelson perceived the complexity of knowledge and our limitations in expressing it early in his life and he has struggled to realize this vision ever since.

> It was an experience of water and interconnection [...] I was trailing my hand in the water and I thought about how the water was moving around my fingers, opening on one side and closing on the

other. And that changing system of relationships where everything was kind of a similar and kind of the same, and yet different. That was so difficult to visualize and express, and just generalizing that to the entire universe that the world is a system of ever-changing relationships and structures struck me as a vast truth. Which it is.

So, interconnection and expressing that interconnection has been the center of all my thinking. And all my computer work has been about expressing and representing and showing interconnection among writings, especially. *And writing is the process of reducing a tapestry of interconnections to a narrow sequence. And this is in a sense elicit. This is a wrongful compression of what should spread out.* (Nelson quoted in Werner Herzog's "Lo and Behold" (2016), *Emphasis added*)

Decades before, in 1970, Nelson argued that "We do not make important decisions, we should not make delicate decisions, serially and irreversibly. Rather, the power of the computer display (and its computing and filing support) must be so crafted that we may develop alternatives, spin out their complications and interrelationships, and visualize these upon a screen." (Nelson 1974, 2003, pp. 332) He is arguing that technology should allow us to design, to see how relationships play out, and to experiment iteratively instead of fixing us onto a serial path. Our emphasis on tool building must focus on creating Thinkertoys to support learning and experimentation, not be used to cement fixed paths and relationships.

Some of the facilities that such systems must have include the following:

- **Annotations** to anything, to any remove
- **Alternatives** of decision, design, writing, theory
- **Unlinked or irregular pieces,** hanging as the user wishes
- **Multicoupling,** or complex linkage, between alternatives, annotations or whatever
- **Historical filing** of the user's actions, including each addition and modification, and possibly the viewing actions that preceding them
- **Frozen moments and versions,** which the user may hold as memorable for his thinking
- **Evolutionary coupling,** where the correspondences between evolving versions are automatically maintained, and their differences or relations easily annotated

(Nelson 1974, 2003, p. 332)

What Nelson is describing here is a dynamic system of relativistic relationships, not a fixed indexing system. The water may close back around your fingers, but it is important to understand where you are and have been in the sea of information. It is easy to get lost in the sea of knowledge. The metaphor of "drowning in information" is completely à propos in this context. Similar to Einstein's thinking, Bush, Engelbart and Nelson are not rejecting the principles of the rationalist Newtonian model. Rather, they argue it is no longer sufficient in organizing and explaining our world. Computers are a tool for taking over the mundane aspects of creating navigation systems to help us better see relationships. They can provide compasses and waypoints on the vast, relativistic sea of information they can store. It is up to us to perceive the relationships that then become apparent.

For centuries, we have had such tools at our disposal, but they were reserved to a narrow spectrum of artists who had the technical skill to execute them. Art is about the creation of new connections in the world, whether that is in the assemblage of tonal elements into a new form of music or the assemblage of visual elements into paintings, photographs, or films.

Computers have significantly lowered the technical barriers to executing thinking on this level. Both Engelbart and Nelson seem to point us toward using digital technology for opening up novel ways of connecting knowledge in ways previously reserved to artists. Nelson has struggled for decades trying to realize his vision through his Xanadu Project, but he recognizes that at its root, even Xanadu is about connecting *textual* ideas. Realized to its fullest potential, a Xanadu project for the 21st Century would eliminate the word "text" from that statement and focus on our ability to connect "ideas." (See **www.xanadu.net**)

Conclusion: Visual Thinking, Relational Thinking and a Concept for Deep Thought

Engelbart gave us many clues about how to use technology to shape thought in "Augmenting Human Intellect" in 1962. One was the thought diagram illustrated previously. As opposed to a set of concepts fixed in a linear textual arrangement, a series of fluid relationships connected his concepts. The idea of concept mapping goes back to the 1930s, but technical limitations (it was obviously not a fluid arrangement) limited it to illustrative purposes. Modern toolsets allow us to approach the dynamic kinds of relationships that Bush, Engelbart, and Nelson seem to have in mind, at least in form if not in substance.

All knowledge is a process of interchange between thought and understanding, whether that is an internal process or the product of collaboration with others. There are geniuses throughout history that have instinctually combined many of these elements internally. However, as thinkers such as Steven Johnson, Frans Johansson, and others have shown, all ideas are a product of social processes as much as they are of individual inspiration. Even ignoring the social aspects of knowledge creation, we can no longer rely on geniuses to make those unexpected leaps of imagination and connections for us. The challenges that we face today are too great and varied for even the most brilliant among us to reliably confront and overcome.

It is beyond the scope of this chapter to explore that fully. However, what is clear is that the process of turning thoughts into understanding is not always a simple one. We have a large variety of tools today that can help us put that into context, such as the simple concept mapping tool I used to create this diagram.

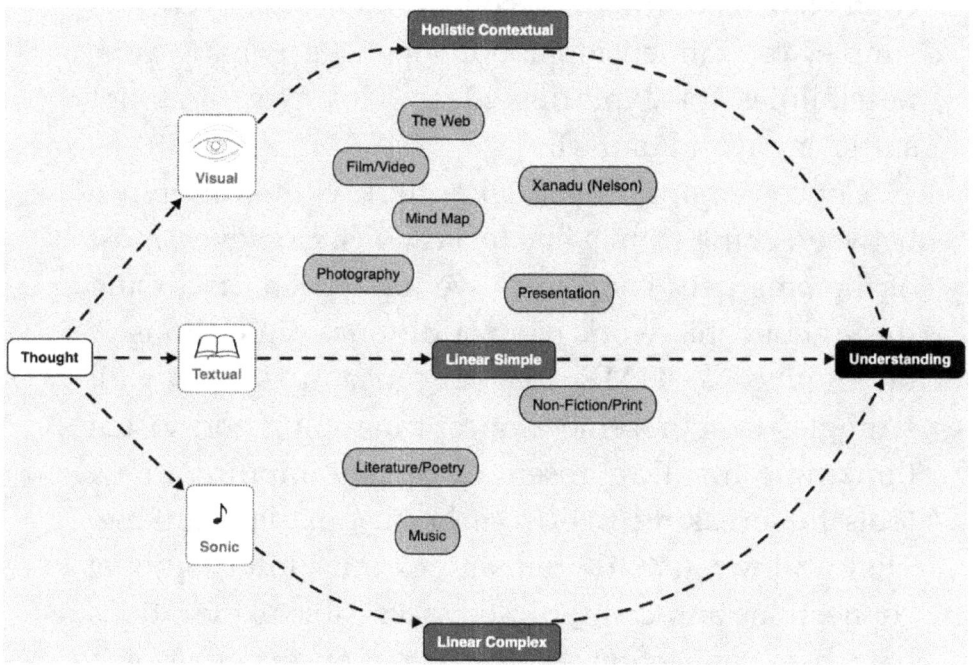

From Thought to Understanding
An Example of a Simple Concept Map

It is interesting that you can see a train of concept mapping throughout the work of Bush, Engelbart, and Nelson. What all three thinkers are essentially suggesting this that we may need to invent a whole new language that effectively combines text and visual elements with a language of connections. Concept mapping gets us there in form, if not in substance. It is a way to leverage the dynamic mapping of visual representations of ideas to create new perceptions of problem sets. I've used it for years in brainstorming sessions and have recently introduced it to my pedagogical repertoire. It adds a second dimension by lifting textual bits of information out of their linear context and places them into a manipulatable graphic context, which adds a second dimension. Even this simple, one-dimensional hop opens up many new possibilities for exploring ideas. However, it is not sufficient in and of itself.

Concept mapping lacks depth. It is the difference between going from a line to a canvas. However, most of the emergent challenges we have been discussing throughout this work have additional dimensions of depth and ⧗ **TIME** that a conceptual map still struggles to represent.[§] Concept mapping and visual thinking can still represent extremely important new tools for breaking free from linear, textual thinking. They allow us to reimagine our informational connections and can represent accessible explorations of those connections. However, concept mapping usually cannot document or capture the intervening journey of how those connections were formed, which is a key element of the vision originally articulated by Bush in 1945. For this, we need to build something far more conceptually ambitious.

[§] Virtual reality might be a possible tool capable of providing a bridge for visualizing data in a dynamic concept-mapping environment. An interesting first step might be the Noda Project (noda.io). However, the step of seamlessly linking this to a robust information network is still missing here.

As is often the case, science fiction may provide us with an important conceptual insight. Douglas Adams created the supercomputer Deep Thought as a key plot point to his series, *The Hitchhiker's Guide to the Galaxy*. In the novel Deep Thought was created to provide the answer to "life, the universe, and everything." After gestating on the answer for 7 ½ million years, the computer responded with the enigmatic "answer" of "42." When its creators expressed consternation over the answer, it replied, "I think the problem, to be quite honest with you, is that you've never really known what the question is." (Adams 1989, p. 121)

The entire *Hitchhiker's Guide* saga is spent searching for an elusive question as if it were an answer. This is a perfect metaphor for the open-ended kinds of questions we need to harness our knowledge network to engage, not answer. There is no ultimate answer to democracy, climate change, poverty, or augmenting human intellect. These are questions that will challenge humanity until the next great paradigmatic shift and beyond. We need to develop systems to manage our engagement with these questions. Our Newtonian knowledge systems are not up to the task for they seek answers instead of questions. This does not mean that we should stop seeking answers but that we have to expand our mental universe to accept that some questions, like that of "life, the universe, and everything," simply do not have answers.

This is a level of abstraction that is at least two steps removed from magical thinking, which is often our natural state. Magical thinking relies on a faith in one set of answers. Rational thinking relies on faith in a fixed set of rules that will lead us to answers. Relativistic thinking plays with those rules to perceive entirely novel sets of solutions, but accepts that these may not represent final "answers" and considers the possibilities that the rules themselves may need to be questioned. Our intellectual tools need to be adapted to that purpose. We made tools to overcome gravity and fly. Now we need tools to overcome our mental limitations and accept ambiguity and an evolving, emergent future.

A 21st Century Deep Thought system would allow us to explore ideas more dynamically with a vast amount of data at our disposal. It's a system designed to facilitate connections between varied and abstract pieces of data. Once again, Engelbart provides us with a vision of how we might do this. In the 1980s and 90s, Douglas Engelbart was not working on improved versions of the mouse. His ongoing frustration with people's inability to grasp the deeper purpose of the work he was doing at SRI in the 1960s led him to organizational thinking. (Rheingold 2000) As part of his "Bootstrap Strategy," now carried on by his daughter Christina at the Engelbart Institute, he developed the concept of "Networked Improvement Communities."

> An improvement community that puts special attention on how it can be dramatically more effective at solving important problems, boosting its

collective IQ by employing better and better tools and practices in innovative ways, is a networked improvement community (NIC).

If you consider how quickly and dramatically the world is changing, and the increasing complexity and urgency of the problems we face in our communities, organizations, institutions, and planet, you can see that our most urgent task is to turn ICs into NICs.

(www.dougengelbart.org/content/view/191/268/)

We already have semi-functioning Networked Improvement Communities, but they lack critical conceptual and communication tools. At a local level, as Engelbart points out, Improvement Communities have functioned for centuries. Steven Johnson and others have pointed out that the coffeehouses of Europe, particularly England, were central to the incubation of the Enlightenment. (Johnson, 2010) Expanding this on a global scale, however, remains an ongoing challenge. Networked Improvement Communities, such as Arizona State University's ShapingEDU Project, struggle to overcome centripetal forces that pull its members back to localized concerns.

Turning ICs into NICs requires a complementary technology, the Dynamic Knowledge Repository. The Engelbart Institute describes DKRs as:

A dynamic knowledge repository is a living, breathing, rapidly evolving repository of all the stuff accumulating moment to moment throughout the life of a project or pursuit. This would include successive drafts and commentary leading up to more polished versions of a given document, brainstorming and conceptual design notes, design rationale, work lists, contact info, all the email and meeting notes, research intelligence collected and commented on, emerging issues, timelines, etc.

(**www.dougengelbart.org/content/view/190/163/**)

We have vast databases that store and index data, but these are often technically divorced from human communities. The tools are simply not there to create an ongoing and dynamic relationship between the information and network. This is because those databases are not Dynamic Knowledge Repositories. They incorporate text-based indexing systems that Bush rejected as inadequate more than 75 years ago. As a result, modern databases do not effectively "record exchanges," as Engelbart put it, at least not in the sense of knowledge. Like many technologies today, we miss that critical *connection* between the vast amounts of information stored in databases and the human processes needed to connect that information into actionable knowledge.

Deep Thought is conceptually a "Dynamic Knowledge Repository." At its root, it would *connect* data to form fresh sets of questions, and therefore knowledge. Using principles of "Emergent Design" as recently described in Ann Pendleton-Jullian and John Seely Brown's excellent, *Design Unbound* (MIT Press, 2018), the project would create a wide range of tools based on visual mapping to create a truly dynamic database tool designed to support a wide variety of Networked Improvement Communities. We can explicitly design these systems to grow and change dynamically as circumstances demand. Tools such as Augmented Reality, Virtual Reality, and blockchain (for authentication, not sequestering information) would form valuable adjuncts in helping us to visualize knowledge in completely new and dynamic ways. Instead of replacing human thought processes, Artificial Intelligence would be employed to guide us to potentially interesting patterns in connections and knowledge that live in the intersections between traditional disciplines.

The Deep Thought concept can trace its roots back to Bush via Engelbart and Nelson. However, it is also deeply rooted in the history of ideas. A flurry of intellectual activity sparked the Enlightenment in the late 17th Century in Europe through the network of ideas embodied in "The Republic of Letters." Its correspondence formed a Dynamic Knowledge Repository for the Networked Improvement Community created by its scientists and thinkers. This networked community helped to plot a way out of the religious wars brought about in part by the invention of the printing press two centuries earlier.

> "[The Republic of Letters] was an institution perfectly adapted to disruptive change of unprecedented proportions. Its history raises the question of why Europe's diversity brought progress in scholarship when by rights disunity

ought to have crippled it. The answer is as simple as it is radical: with existing institutions of learning in crisis or collapse, the Republic of Letters founded its legitimacy on the production of new knowledge." (McNeely and Wolverton, 2008, p. 123.)

The internet is the printing press of our age. We find ourselves at a new crossroads, lost in a sea of information. Vannevar Bush may have perceived this challenge earlier than most, and Douglas Engelbart and Ted Nelson busied themselves in solving the technical challenges it represented, but we still find ourselves in the place that they warned us about. Society is even more in danger of becoming unstuck from its informational anchors than it was in 1945. When this happened in the 1500s, it resulted in more than a century of bloodshed. We have it in our power to bypass that fate, fast forward 200 years, and create a digital Republic of Letters. By combining the best of what text offers with the new vocabulary of digital tools now at our disposal, we must finally create systems that augment how we really think.

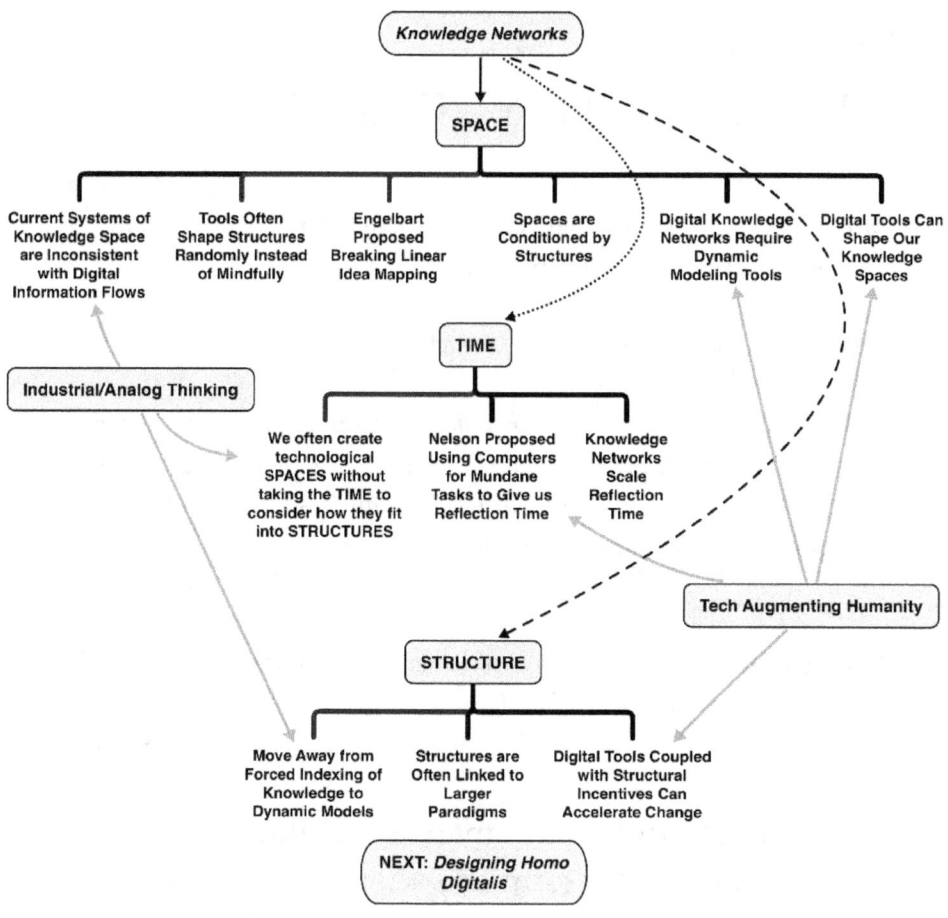

Conclusion
4.0 Designing Homo Digitalis

Evil: When I have the map, I will be free, and the world will be different, because I have understanding.

Robert: Uh, understanding of what, Master?

Evil: Digital watches. And soon I shall have understanding of video cassette recorders and car telephones. And when I have understanding of them, I shall have understanding of computers. And when I have understanding of computers, I shall be the Supreme Being!
Michael Palin/Terry Gilliam, Time Bandits, 1981

We believe that it is technology married with the humanities that yields us the results that make our hearts sing.
Steve Jobs, 2011

We are at our most human as a child. School then conditions us to conform to society. If that society is industrial, it teaches us to conform to the machine. This is a lesson I did not learn because to me the machine was always my servant. I could use it to create worlds. This makes me digital. I am a creator of worlds; worlds of creativity and imagination; worlds of new constellations of ideas; worlds of novel stories to explore and follow. Computers gave me the metaphor of a software world, and they gave me the tools. They did not make me digital. My inner digital child was always there.

We are already *homo digitalis*. It's just locked away inside the years of indoctrination we've had being *homo industrialis*. Learning to become digital is learning to become human; to tell our stories with machines and not have machines dictate our stories. The digital revolution offers us a way to circumvent the imposed linearity of the industrial world, pause, reflect, and play with the worlds we create for ourselves.

In a weird way, as I conclude this book, it brings me back full circle to the dawn of my own technological existence. I have always grounded my vision of technology in a sense of possibility, a sense of removing constraint. Steve Wozniak was an early technological hero to me. He built the original Apple computer because he wanted to solve problems. When he felt like the company that he helped to create, Apple Computer, was no longer a vehicle for him to solve problems, he left, went back to school, and spent the next 20+ years educating and inspiring children to follow in his footsteps in much the same way as he inspired my teenage mind.

Because of people like Woz, I operated with the mindset of Engelbart, Nelson, and Kay long before I knew who any of them were. When I was selling cameras, I served the artists and the art, not the gear. When I was doing tech support in the late 90s, I always understood that I was serving the users and not their machines. This, in my mind, is what separates a technologist from a technician.

Like a child, I have always sought to bend technology to my will in order to create. Woz expresses his creative determinism through his engineering and coding. I express mine through writing, photography, and teaching. Our attitude about technology is, as far as I can tell, the same: it is a liberating force in our lives. Recently, Woz reflected on what excited him about starting Apple in the 70s, "the little geek like me was going to be writing programs and solving problems for our companies and we were going to be more important than the big huge companies." (Wozniak 2014) This is a very humanistic, incorrigible vision of the future: technology is there to help us capture our childhood creativity and turn it into work that benefits, and ultimately serves, others.

The spirit of the 70s that Woz, Nelson, Kay and others of their generation of technologists embody is something we need to recapture. One thing that seems to unite all of them and me is that we struggle to maintain that spirit of the child within us. Digital technologies have given us wonderful tools for expression and, in doing so, have also given us powerful tools for problem-solving. There is joy in that, even though it may have faded in the intervening years. It still represents a vision that we never truly gave up on.

In my approach to technology, I have always sought to help others rediscover their childish perspective. I strongly believe that it is critical to the future of humanity and the human within us to find that childish joy of empowerment again. Personally, I find it most often when I'm teaching and working with groups to brainstorm our way out of problems. I also find it when I am creating art, both visual and textual.

Often, those blend with each other. Collectively, I hope this book shows many people how to design and make real the dreams of technological visionaries of the Seventies.

It is only through finding our "childish" creative spirit that we can redesign the relationship we have with our machines. Changing that relationship allows us to redesign relationships with ourselves, each other, and our ideas. As I have stressed repeatedly in this volume, there is very little new under the sun. We are dealing with the same variables of human perception and culture as Julius Caesar or Machiavelli or Madison. Like a child, we must learn to be humble in the face of our challenges.

Technology may change the speed and frequency of how we exchange data. This information overload can have a profound effect on *how* we interact with one another. However, the purposes and ends of those interactions remain the same as they have for millennia. We should not overlook that as we navigate the blizzard of information and misinformation tossed at us every day. It can provide us with a path through the storm. We just need better tools to help us navigate it. Technology, provided we design it properly, can show us better ways to manage the struggles we have with ourselves and each other.

> [D]igital technologies let us do things differently – from the digital humanities to 3-D printing – and emerging analytical and visualization methodologies let us see differently. Doing and

> seeing differently means that we can begin to interact with and experience the world differently. A new twenty-first century ontology is emerging. (Pendleton-Jullian and Brown 2018, p. 29)

Learning to live with our technology requires learning to become more human. Technology can facilitate human understanding. However, as a teacher and parent, I have realized through hard experience that learning only happens if the learner decides to learn. I am optimistic we can grow to learn if we let ourselves.

As a political scientist, people often ask me what my view of politics is. I respond that I am a "naïve realist." I believe in humanity and what it can become. I have seen the magic that individuals and groups coming together can do. I see this in my students, my children, and my friends. It's in my colleagues from the education, business, art, and government worlds. I am, however, also a realist in understanding how even the most enlightened human systems can fall into traps of behavior that stifle that innovative child in all of us. I agree wholeheartedly with Peter Morville when he writes:

> Code is a function of culture. That's one of the most important lessons I've learned in 20 years of consulting. It's not that the tail can't ever wag the dog, but when it does, it usually happens quite slowly…. If we fight culture, it will fight back and usually win. But if we look deeper, and if we're open to changing ourselves, we may see how culture can help. (Morville, 2014)

The social scientist and historian in me also perceives the mental roadblocks we put in front of us, from dehumanizing systems to mass delusion to learned helplessness. This book is itself an act of naïve realism. It attempts to grasp at possibilities that Engelbart, Nelson, Kay, Bush, and many others perceived in a future augmented by rapidly-evolving technology. As Ted Nelson said in his eulogy to Douglas Engelbart:

> I used to have a high view of human potential. But no one ever had such a soaring view of human potential as Douglas Carl Engelbart — and he gave us wings to soar with him, though his mind flew on ahead, where few could see. (Nelson 2013)

The book presents this kind of world of possibilities, while recognizing its perils if we proceed without mindful reflection to fight for our humanity. As Nelson subsequently says, "Perhaps his notion of accelerating collaboration and cooperation was a pipe dream in this dirty world of organizational politics, jockeying and backstabbing and euphemizing evil." (Nelson 2013) I dare to hope that technology can help us find our way out of our societal morasses by exposing outdated industrial paradigms that make them seem intractable.

The realities that fear, habit, and perceptual blinders put on us individually and collectively are very real to me. There are pathways out of that morass of inhumanity. We really only have two choices: we can use technology to make the world human again, or we can use it to perpetuate the inhumane legacies of the last two centuries. It was my goal in this book to make this choice clear. It is possible to create healthier relationships between ourselves and our tools.

There are many counter-narratives about what I am trying to get across here. The media today is flooded with dark pronouncements about our technology going awry. In some ways, this narrative echoes the wild pronouncements that marked the early years of the Atomic Age in the 50s and 60s. Lurid futures make for popular media from *Godzilla* to *Planet of the Apes* to *The Matrix*. It should come as no surprise, therefore, when the news media picks up on these cues and quickly brands new technological developments as the coming of the apocalypse.

We should never underestimate technologies like atomic power for the dangers that they hold. However, these dangers need to be understood from sober and profoundly human perspective. At the literal dawn of the atomic age, Robert Oppenheimer related a passage of Hindu scripture, "Now I am become death, destroyer of worlds." In using this phrase, he immediately grasped our centrality in the process of technology. The passage didn't say, "The bomb is become death." It says, "*I* am become death." By using this phrase, Oppenheimer explicitly acknowledged the centrality of the human in the technological equation. He understood implicitly that we are intimately bound to our tools. They are our responsibility.

Technology is amoral. It is in our failure to recognize that reality that we face the greatest dangers created by human activity. Oppenheimer recognized this reality. We collectively and individually miss it as we constantly look past ourselves and just see the technology. Ironically, it may be the invisible danger of carbon emissions that pose our greatest existential threat, not radioactive slag or killer machines. That threat is entirely the product of human decisions, not our technology.

Worrying about radioactive slag and killer machines, while occasionally making for good (and bad) cinema, is not really a productive way to think about technology. To do that, we need to look in the mirror. Our technology is us. A clear and sober assessment of any new technology, weighing its costs and benefits to ourselves and our society, should be the starting point of any discussions about purchases and implementation. Positively ("naïvely") augmenting human intellect, and, by extension, humanity itself, should always be our goal. A sense of play (hacking) should guide us toward experimentation. Technology should open those doors, not close them.

We need systems of alternative, safe energy generation to replace the fossil fuels that are warming our atmosphere. We will only develop them through play. Instead of risking radiation poisoning, or worse, we can model new types of nuclear reactors using simulations. Putting things in the air is essential for modern transportation and communication. Flying efficiently is critical to combating climate change. Now, with the help of computing, we can model aircraft into ever more efficient shapes without risking pilots' lives. These examples have one thing in common: they look past the means and towards the ends and creatively use technology to achieve those ends, not "better" technology. Iteration means iteration toward the better, not simply an improvement over last year's model.

Managing the challenges of the digital transition requires a systematic approach to change. I am under no illusions that this is in any way a trivial task. One of the ongoing and most difficult challenges of the West Houston Institute is making it complement the existing system at Houston Community College, as well as higher education generally. The WHI example discussed in Chapter 3.2 is illustrative of any attempt to insert a digital ❋ **STRUCTURE** into an analog system. Analog systems don't handle failure or iteration particularly well. When the tape breaks, it stops the story. However, failure is built into the process of any digital system. Everything in the WHI was designed to be lightweight and adaptable. It is what the technology *enables*, not what it *is* that matters.

Analog systems demand that we establish a linear logic up front. They assume all efforts require a massive tail to make them work. If you build a machine shop as part of a Makerspace, you better be teaching machine shop workforce classes there. Instead of looking it at as a flexible tool, the system views the shop as a fixed asset tied to a fixed set of programmatic activities. Enrollment in those programs is a measure of success. Otherwise, you can't have the ✿ **SPACE**. Instead of supporting human augmentation, everything you do must support the logic of the system.

Systems are sets of stories we tell ourselves. The dog usually wags its tail, not vice versa. However, there are many ways that we can now use digital tools to "do and see differently." Donella Meadows's leverage points show that a common theme in most human challenges lies in how we perceive opportunities and constraints. (Meadows 1999) Technology can help us see opportunities instead of barriers, but only if we design it with hacking systemic barriers in mind.

One of our human challenges is seeing narrative pathways to opportunities. We must learn to minimize the drag of constraints by not being blinded by what we perceive as "the possible." The COVID-19 Pandemic can shine a light on the unexpected digital opportunities that emerged under difficult circumstances, or we can brush aside its lessons as an aberration. If we only see constraints, we will inevitably regress back to what we accepted as "normal." If we see opportunities, we can reshape how our societies work for the better. Growth and change require that we cannot be afraid to develop new narratives.

How we perceive economics is one illustration of how these perceptions can fundamentally color and shape narratives. A case in point is one of the other technological bogeymen that has emerged over the last decade: the argument that humans are being replaced by machines.

Economists Erik Brynolfsson and Andrew McAfee fired the first warning shots with their 2011 book, *Race Against the Machine* (Brynolfsson and McAfee 2011) and followed this up with *The Second Machine Age.* (Brynolfsson and McAfee 2014) Other studies, such as those by Oxford University economists Carl Benedikt Frey and Michael Osborne (Frey and Osborne 2013) and an OECD Study from 2018 (OECD 2018) argued that between 15 and 47% of all jobs in the developed world will be threatened by automation by the mid-2030s. These economic analyses got the attention of the media and others and led to a lot of hand-wringing over the future of humans as part of the economy.

The machines are coming for us is an old story. An Industrial Age analogy is the battle between John Henry and the steam pile driver. I admit that I have used this one myself to knock people out of their complacency. Framing is everything here. Neither Brynolfsson and McAfee nor the authors of the Oxford Study are wrong. However, as they themselves point out, the problem lies not with technological advances, but in how humans approach it. If we are masters, automation will be extremely empowering, liberating us from dangerous, repetitive, dehumanizing tasks. If we are slaves, it threatens our stability and place in the world.

How we see the world is the key distinction between whether we decide to become masters or remain slaves. It is our choice. As the Oxford study notes, "making predictions about technological progress is notoriously difficult." (Frey and Osborne 2013, p. 47) This is because "technological" progress is a product of the symbiotic relationship we have as individuals and societies with our machines. Hearkening back to Oppenheimer, the nature of our machines is up to us, not them. The reality is that machines will not replace humans anytime soon.

Brynolfsson and McAfee point out that the most effective implementations of technology involve humans working closely with their machines. They cite the example of how international chess has evolved since the defeat of Garry Kasparov by Deep Blue in 1997. Kasparov himself relates how the most effective chess players have evolved into teams of humans working closely with computers to develop, test, iterate, and execute strategies to beat other similar teams, with the creative element of *how* the technology is being used to solve the problem of winning a chess game becoming the determining factor in who wins. (Kasparov 2010) If you think about it, most great art is humans fusing with technologies, whether represented by a brush, violin, or a camera. As an artist, I am at my best when I use my camera and digital workflow as a mechanism for extending my imagination.

Effectively integrating humans with machines continues to be one of the most difficult and pressing problems within our society. Putting obstructions, through bad interface design or difficulty in even accessing tools, between humans and their technology is the most significant barrier to using our technology to augment our humanity. Removing these barriers must be a priority if we want to grow our societies and educate our children to face a future burdened with the legacies of two centuries of industrialization.

If we persist in a mindset that looks on machines as adversaries instead of partners, replacement is the more likely outcome than augmentation. The technology environments we create in this scenario will be inherently hostile to positive human interaction except through a technological elite. It would mark an unprecedented consolidation of power in the hands of the few. This is not a healthy future for humanity.

It is every bit as possible to construct technology to liberate us, as it is to design it to take away our freedom and humanity. Both approaches will move humanity forward, but only one represents genuine progress. This is the reason that we need to demand the former course from our technologies, both large and small, and boycott those who seek to monopolize technological resources. Technological hierarchy is not an inevitability. It is a choice.

We must recognize that the primary barriers here are not technological limits but societal, cultural, and normative choices that we make as humans. To cite just one example, the reason we don't have universal broadband is because of political and economic choices, not technological barriers. Yet we persist in accepting this "normality" because technology is supposed to be difficult and hard to access. Our industrial relationship with machines was inherently adversarial. Machines were often complex and required specialists to control. This thinking has spilled over into our relationship with computing technology. We find it reflected by the doomsayers that predict machines will supplant human activity rather than augment it. It is up to us to design our way into a better set of relationships.

I have hope for a better technological future because I have seen how my existence has been enriched and augmented by my relationship with the machines around me. At their best, my machines have helped me to find the human in myself. They allow me to share myself with others through photography, writing, and many forms of visual storytelling (concept/mind mapping, slides, web design, etc.). I have seen time and time again how humans always seem to carve a human element into any set of technologies, whether those are "impersonal" computers or clinically designed physical spaces. It's just a question of how hard this is and whether the human is disintermediated so much that the normal consequences for antisocial behavior are suppressed (or stressed).

One thing that gives me hope about Digital Age technology is that it is far more subject to human hacking than Industrial Age technology ever was. It was almost impossible to hack an industrial steel mill, but you can hack Twitter today. Even in the Industrial Age there were hackers such as those who took "stock" cars and "hot-rodded" or otherwise modified them to suit their individuality. Nowadays it's much easier (and cheaper) to "hot-rod" a WordPress site to make it yours and to adapt it to very specific purposes.

We have created communities of sharing that help provide many of the more difficult building blocks. These communities distribute those tools willingly as a service to each other in much the same way that the hotrod enthusiasts shared tips and tricks for generating more power from their engines. What technology has done has been to democratize a wide swathe of tools. These tools have the capacity to turn all of us into creators and hackers of reality. We all need to internalize the ideals that Steve Wozniak had as he was pursuing his vision of personal computers back in the 1970s.

We must design technology to support a network of ideas instead of allowing it to channel those ideas for the profit of the few. Innovation comes from within a community of ideas. As Woz said in his 2014 talk, "minds are more important to me." (Wozniak 2014) All technology design therefore needs to work to bring humans together and to provide for the free interplay of ideas. This is as true in a company as in a school.

In order to preserve our ability to hack and redesign the systems around us, a set of basic principles needs to be kept in mind at all times. Throwing up walls around ideas stifles our growth as humans, individually and collectively. The Digital Age offers us the possibility of knocking down these walls. In the 1970s the dream was that "we will be masters of our own destiny." (Wozniak 2014)

It's no accident that the tool Wozniak created, the original Apple computer, launched my technological journey. That journey has ultimately led me to type these words (and many other things along the way). Along the way, I have learned to embrace 5 principles for a healthy relationship with technology. In no particular order, they are:

1. Don't be afraid to hack and play with technology
2. Use technology to hack and play with your environment
3. Use technology to hack and play with your ideas
4. Use technology to hack and play with others
5. Transparency is the key element in making technology a force for good

These five principles are crucial in making the IdeaSpaces framework functional. They are central to any holistic design for solving problems. We need to recognize that everything is pliable to design effective digital ✪ SPACE, both in its physical and virtual forms. We need to recognize that ideas are the product of networks of humans and that we need to design digital technology to share and hack those ideas. Humans require ⧗ TIME to hack themselves, their ideas, and their technology but technology expands the number of avenues that can be explored exponentially. Finally, we must recognize that humans create vast and complex ✱ **STRUCTURES** to make sense of, scale, and order their realities. The tools made available by the Digital Age threaten many existing ✱ **STRUCTURES**, just as Industrial Age ✱ **STRUCTURES** threatened agrarian ones two centuries ago. It is possible to design ✱ **STRUCTURES** that facilitate human creativity and innovation rather than retard change. For any significant effort at implementing and augmenting human intellect to work, all three levels must be a part of the design plan.

I keeping coming back to the centrality of transparency in addressing the central challenges of digital transformation. Without transparency, it is impossible to adopt any sort of holistic design strategy. Ignorance of any significant part of the technological, human, or structural elements that change requires may cause the entire project to be threatened, compromised, or even stillborn.

I remain optimistic because Digital Age technology favors transparency over the long term. I refer to John Gilmore again, "The internet interprets censorship as damage and routes around it." This is still true and is the root of my faith in technology to provide transparency. Transparency allows for design; it allows for augmentation; and it is essential for positively hacking the world. We have a better chance of understanding ourselves and our societies than at any other time in human history because of digital technology. We just have to grasp it. For that, we need to design human-centered machines, not machine-centered humans. The right tools are essential so that we can mindfully engage challenges and take full advantage of that transparency.

We must also recognize that to find ourselves again and adapt to *homo digitalis* requires that we move at a profoundly human timescale. This will require small, iterative steps down paths that are not predetermined but rooted in a clear-eyed, transparent view of the world. The urgency of those problems may dictate that we explore many trails quickly, but, again, technology allows us to do that in unprecedented ways. We just have to keep trying doors until one of them leads us into a better future.

I leave you with the words of Ted Nelson, a true internet dreamer, who said over 40 years ago that, "every change is merely a small difference in degree." (Nelson 1979) Let us recognize that, pick up our tools, and start making the small differences that will lead to a better future for us and our children. I hope this book has given you some ideas about how you can do that.

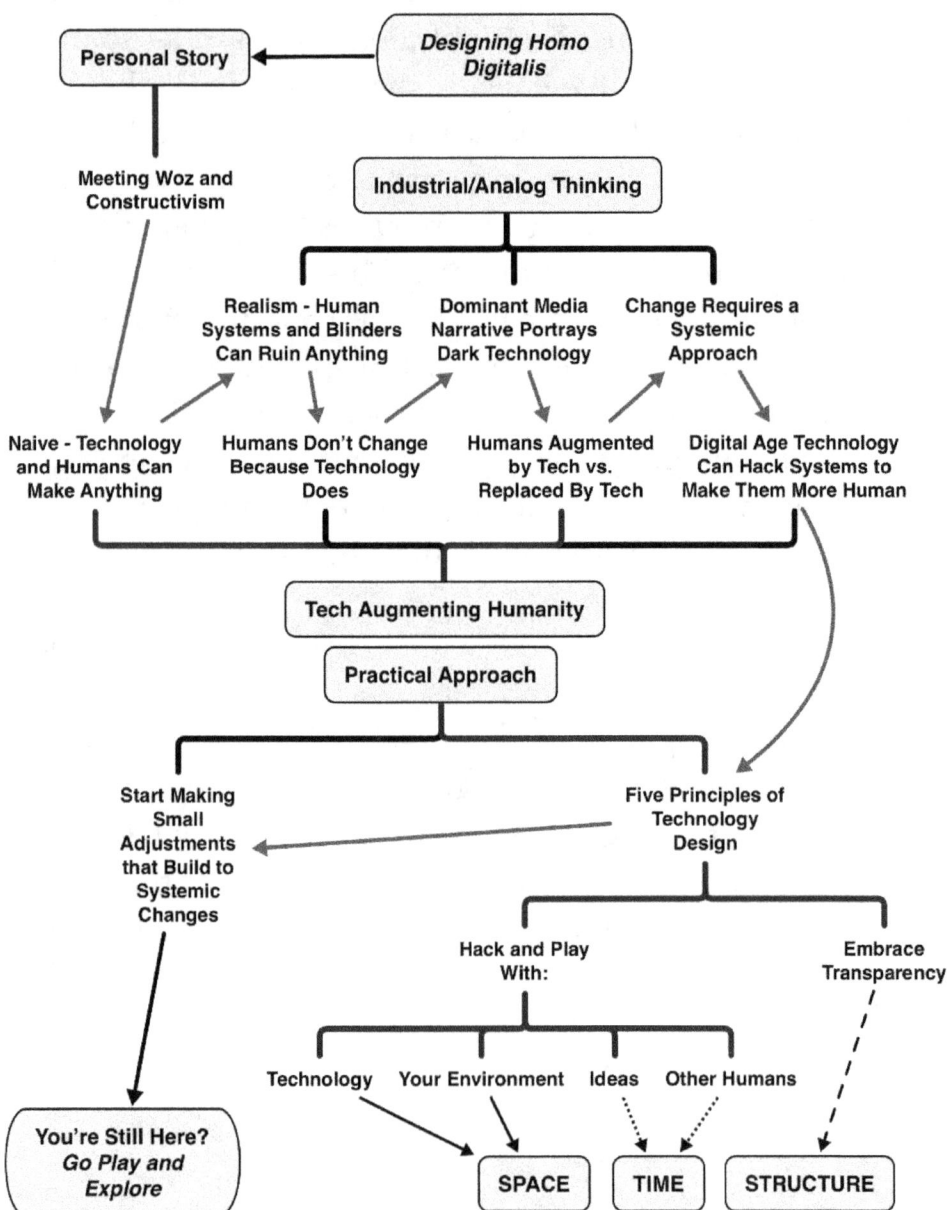

Acknowledgements

There are many people who have contributed to this book over the years it has taken to produce. I am not immune to the IdeaSpaces frameworks and have depended on my own networks of ❋ **STRUCTURES** in developing the thoughts expressed in this book.

I have had the privilege of working with many talented people over the years. In 2017 and 2018, I was privileged to converse regularly with Gardner Campbell, Ruben Puentedura, Bryan Alexander, and David Wedaman in an informal reading circle. Some very important ideas in this book emerged from these discussions and readings.

Over the course of the last two years, I have worked closely with a group of thinkers in the ShapingEDU Project from Arizona State University, ably led by Lev Gonick and Samantha Becker. Lev and Sam have provided much support over the years and the group of "mayors" I have been lucky to work with include Jonathan Nalder, Paul Signorelli, Laura Geringer, Lisa Gustinelli, Mark Fink, Emory Craig, Maya Georgieva, Nancy Rubin, and Tanya Joosten who constantly challenge and augment my ideas. My fellow Storytellers in Residence, Paul Signorelli and Karina Branson, have contributed in untold ways to my skills in telling the stories of the Digital Age.

I had many rich discussions with the creative team at the West Houston Institute as we developed the building and have subsequently sought to make it operational. Edmund "Butch" Herod ably led this team, and it included Jordan Carswell, Susan Thompson, Alexandra Almestica, and Laura Williamson, all of whom have contributed to the thinking in this book in significant and unique ways. Zachary Hodges, president of Houston Community College's Northwest College, also has my deep gratitude for letting me/us "play" with a $50 million building project.

I want to single out for special thanks Bryan Alexander and Ruben Puentedura, who have become close friends over the years and have offered deep and ongoing support and thought that contributed to the creation of this book.

I also want to single out those who have generously read various drafts of this book. Ruben Puentedura, Paul Signorelli, Butch Herod, Brian Murfin, Carter Haymes, and Wini Haymes have all struggled through my prose and found many mistakes and inconsistencies that I have endeavored to correct.

My family has been patient and understanding of the long hours and sacrifices that have been required of me. For this, I am particularly grateful to my wonderful wife, Lorie, and to my children. My parents, Ed and Wini, have always offered me unwavering support. This book would not have happened without them.

Finally, I want to thank my ever-patient publisher and editor, Jared Bendis, who has helped keep me on course through an often-meandering process over the last couple of years.

A Special Afterword on the Pandemic and Digital Humanization

I wrote most of this book prior to the outbreak of the Covid pandemic of 2020. As the world moved suddenly into remote environments, we suddenly leaned on our digital tools to an extent that most found profoundly dehumanizing. This is not surprising because at the same time many of us moved into very reactive positions within our organizations and habits (❇ **STRUCTURES**) as we desperately struggled to grasp at "normalcy." It was sadly predictable to see that most reached backwards instead of forwards as we fumbled our way toward responses at both the systemic and individual levels.

The crisis I perceived in education was so great that I pivoted into writing a second book before I fully completed this volume. I based that book, *Learn At Your Own Risk*, on principles we have explored in this volume. I always intended the ideas in this volume to explore the fundamental nature of our relationships, collectively and individually, to technology. The crisis laid bare the weaknesses in those relationships, like many other connections in our societies. However, the principles that underlie effective adaptation to a digital existence did not shift. Therefore, other than a few examples here and there throughout the text which came up during the crisis and illustrate some points more fully, I have not felt compelled to engage in any comprehensive rewriting of the foundations of the text.

I continue to engage in observations of individuals' and institutions' reactions to the crisis. Perhaps these will form chapter addenda to a second edition, or perhaps they will form the core of an entirely new volume. I always intended *Discovering Digital Humanity* to form the beginning of a series of conversations about what it means to be digitally human. *Learn at Your Own Risk* represents one such effort; one specifically directed at teachers. I have also begun a volume focused on the needs of decision makers in the private sector based on work done for this volume that represents a longer-term exploration of how digital transformation will impact our organizations, how they work, and the systemic structures that support them.

This conversation will not end in my lifetime. Like all ideas in complex societies, our relationship with technology will be subject to constant renegotiation. I hope that this volume contributes in some small way to these conversations as we recover from the pandemic. I hope it will form a part of the much more complex discussions beyond that as we grapple with the twin challenges of automation and climate change, both of which will require us to adapt and grasp at unimagined opportunities that will dwarf our present circumstances.

Tom Haymes
Katy, Texas, March 2021

End Notes

0.1 Introduction

Raworth, Kate, *Doughnut Economics: 7 Ways to Think Like a 21st Century Economist* (White River Junction, VT: Chelsea Green, 2017).

Johnson, Steven, *Where Good Ideas Come From* (New York: Penguin, 2010).

Chapter 0.2

Brown, Frederic, "Answer" in *Angels and Spaceships* (New York: Dutton, 1954) also at:
www.freesfonline.de/authors/Fredric_Brown.html

Rheingold, Howard, *Tools for Thought*, 2nd ed. (Cambridge, MA: MIT Press, 2000).

Jarvis, Jeff, *Public Parts: How Sharing in the Digital Age Improves the Way We Work and Live*, (New York: Simon and Schuster, 2011).

Kuhn, Thomas, *The Structure of Scientific Revolutions*, 3rd ed. (Chicago: University of Chicago Press, 1962).

Chapter 1.1

Licklider, J. C. R, "Man-Computer Symbiosis" in Noah Wardrip-Fruin and Nick Montfort eds., *The New Media Reader*, (Cambridge, MA, MIT Press, 2003), pp. 73-82 also available at: archive.org/details/TheNewMediaReader

Csikszentmihalyi, Mihaly, *Flow: The Psychology of Optimal Experience* (New York: Harper Collins, 1990, 2008).

Kay, Alan, "Doing with Images Makes Symbols," at: archive.org/details/AlanKeyD1987_2

Engelbart, Douglas, "Augmenting Human Intellect," in Noah Wardrip-Fruin and Nick Montfort eds., *The New Media Reader*, (Cambridge, MA, MIT Press, 2003) pp. 93-108 also available at: archive.org/details/TheNewMediaReader

Engelbart, Douglas, Richard W. Watson and James C. Norton, "The Augmented Knowledge Workshop," in AFIPS Conference Proceedings, Vol. 42 (1973), pp. 9-21 available at: www.dougengelbart.org/content/view/133/000/

Shannon, Claude, "A Mathematical Theory of Communication," *The Bell System Technical Journal*, Vol. 27, pp. 379–423, 623–656, July, October 1948 available at: affect-reason-utility.com/1301/4/shannon1948.pdf

Nelson, Theodore, *Dream Machines* (excerpt) in Noah Wardrip-Fruin and Nick Montfort eds., *The New Media Reader*, (Cambridge, MA, MIT Press, 2003) p. 301-338 also available at: archive.org/details/TheNewMediaReader

Allen, Doug and Daniel Castro, "Why So Sad? A Look at the Change in Tone of Technology Reporting from 1986 to 2013." *Information Technology and Innovation Foundation* (Washington, February 2017) at: **www2.itif.org/2017-why-so-sad.pdf**

Chapter 1.2

O'Carroll, Aileen, *Working Time, Knowledge Work and Post-Industrial Society*, (New York: Palgrave Macmillan, 2014).

Mechanical Turk at: **www.mturk.com**

Houston, Rab and K.D.M. Snell, "Proto-Industrialization? Cottage Industry, Social Change, and Industrial Revolution," *The Historical Journal*, Vol. 27, No. 2 (Jun., 1984), pp. 473-492.

Csikszentmihalyi, Mihaly, 1990, 2008, *Ibid*.

McCrossen, Alexis, *Marking Modern Times*. (Chicago: University of Chicago Press. 2013), Kindle Edition.

Munro, Allen, "Alan Kay Thinks the Computer You Just Bought Is 'No Big Deal'" *St. Mac*, Volume 1/3 (April 1984) at: **archive.org/stream/st.mac-1984-04/st.mac-1984-apr**

Canales, Jimena, *The Physicist and the Philosopher: Einstein, Bergson, and the Debate that Changed Our Understanding of Time*, (Princeton: Princeton University Press, 2015), Kindle Edition.

Sprague, Shawn "What Can Labor Productivity Tell Us About the U.S. Economy," *Beyond the Numbers*, v. 3/12 (May 2014), US

Bureau of Labor Statistics at: www.bls.gov/opub/btn/volume-3/what-can-labor-productivity-tell-us-about-the-us-economy.htm

McCrossen, *Ibid.*

Haymes, Tom, *Learn at Your Own Risk: Nine Strategies for Thriving in a Pandemic and Beyond* (Cleveland: ATBOSH Media Ltd., 2020).

Pink, Daniel, *When: The Scientific Secrets of Perfect Timing* (New York: Penguin, 2018).

Isaacson, Walter, *Benjamin Franklin: An American Life* (New York: Simon & Schuster, 2004).

Seppälä, Emma, "Three Science-Based Reasons Vacations Improve Productivity" *Psychology Today,* August 17, 2017 at: **www.psychologytoday.com/us/blog/feeling-it/201708/three-science-based-reasons-vacations-boost-productivity**

Pink, Daniel, "The Flip Manifesto" (2013) at: **www.danpink.com**

Dunbar, Kevin, "How scientists think: On-line creativity and conceptual change in science," in T. B. Ward, S. M. Smith & J. Viad (eds.), *Creative Thought: An Investigation of Conceptual Structures and Processes* (New York: American Psychological Association, 1997), pp. 461-493.

Ries, Eric, *The Lean Startup*, (New York: Crown Business, 2011).

Ann Pendleton-Jullian and John Seely Brown, *Design Unbound* (MIT Press, 2018).

Chapter 1.3

Brand, Stewart, *How Buildings Learn: What Happens After They're Built* (New York: Viking, 1994).

Kelly, Kevin, *What Technology Wants* (New York: Viking, 2011).

Johnson, Steven, 2010, *Ibid*.

Conway, Melvin, "How Do Committees Invent?" *Datamation*, (April 1968), pp. 28-31 also at: **www.melconway.com**

Chapter 2.1

Kirschenbaum, Matthew, *Track Changes: A Literary History of Word Processing* (Cambridge, MA: Harvard University Press, 2016).

Murray, Janet, *Hamlet on the Holodeck* (Cambridge, MA: MIT Press, 1997).

Shannon, 1948, *Ibid*.

Nelson, 1974, *Ibid*.

Kay, Alan, "Programming and Programming Languages," (2010) at: **www.vpri.org/pdf/rn2010001_programm.pdf**

Papert, Seymour, *Mindstorms: Children, Computers, and Powerful Ideas* (New York: Basic Books, 1980, 1993).

Eshet-Alkalai, Yoram and Nitza Geri, "Does the Medium Affect the Message? The Influence of Text Representation Format on Critical Thinking," *Human Systems Management* 26 (2007), pp. 269-279.

Kay, Alan, "Computers, Networks, and Education," *Scientific American*, (September 1991) pp. 138-148

Papert, 1980, *Ibid.*

Nelson, 1974, *Ibid.*

Roberts, Siobhan, "Claude Shannon, The Father of the Information Age Turns 1100100" (April 30, 2016) at: www.newyorker.com/tech/annals-of-technology/claude-shannon-the-father-of-the-information-age-turns-1100100

Shannon, Claude E., *The Mathematical Theory of Communication* (Bloomington, IL: University of Illinois Press, 1963).

Morgan, Michael Cotey, *The Final Act: The Helsinki Accords and the Transformation of the Cold War* (Princeton, NJ: Princeton University Press, 2018).

Elmer-Dewitt, Philip "First Nation in Cyberspace," *Time*, Dec. 6, 1993 at:
kirste.userpage.fu-berlin.de/outerspace/internet-article.html

Von Stackelberg, Peter and Alex McDowell, "What in the World? Storyworlds, Science Fiction, and Futures Studies," *Journal of Futures Studies*, v. 20/2 (December 2015) pp. 25-46.

McCloud, Scott, *Understanding Comics: The Invisible Art* (New York: HarperCollins, 1993).

Meadows, Donella, "Leverage Points: Places to Intervene in a System," (1997) at: **donellameadows.org**

Bush, Vannevar, "As We May Think," in Noah Wardrip-Fruin and Nick Montfort eds., *The New Media Reader*, (Cambridge, MA, MIT Press, 2003) pp. 37-47 also available at: **archive.org/details/TheNewMediaReader**

Cassel, David, "Ted Nelson: What We Can Still Learn from Xanadu" at: **thenewstack.io/ted-nelson-can-still-learn-xanadu/**

Wilford, John Noble, *The Mapmakers* (New York: Random House, 1981).

McLuhan, Marshall, "The Medium is the Message," in Noah Wardrip-Fruin and Nick Montfort eds., *The New Media Reader*, (Cambridge, MA, MIT Press, 2003) pp. 193-210 also available at: **archive.org/details/TheNewMediaReader**

Standage, Tom, *Writing on the Wall: Social Media, The First 2000 Years* (New York: Bloomsbury, 2013).

Tufte, Edward, "Powerpoint is Evil: Power Corrupts, Powerpoint Corrupts Absolutely," *Wired* (September 2003) at: **www.wired.com/2003/09/ppt2/**

Chapter 2.2

Naughton, John, *What You Really Need to Know About the Internet: From Gutenberg to Zuckerberg* (London: Quercus, 2012).

Licklider, J.C.R. and Robert W. Taylor, "The Computer as a Communication Device," *Science and Technology* (April 1968) pp. 21-41 at: **signallake.com/innovation/LickliderApr68.pdf**

Rheingold, Howard, *Net Smart: How to Thrive Online* (Cambridge, MA: MIT Press, 2012).

Wojick, Stefan and Adam Hughes, "Sizing Up Twitter Users," (Pew Research Center, April 24, 2019) at: **www.pewresearch.org/internet/2019/04/24/sizing-up-twitter-users/**

Benkler, Yochai, Casey Tilton, Bruce Etling, Hal Roberts, Justin Clark, Rob Faris, Jonas Kaiser, Carolyn Schmitt, "Mail-In Voter Fraud: Anatomy of Disinformation Campaign" (Berkman Klein Center for Internet and Society at Harvard University, October 1, 2020) at: **cyber.harvard.edu/publication/2020/Mail-in-Voter-Fraud-Disinformation-2020**

Elmer-Dewitt, 1993, *Ibid*.

Stansberry, Kathleen "Experts Optimistic About the Next 50 Years of Digital Life," (Pew Research Center, October 28, 2019) at: **www.pewresearch.org/internet/2019/10/28/experts-optimistic-about-the-next-50-years-of-digital-life/**

Ferguson, Niall, *The Square and the Tower: Networks and Power, from the Freemasons to Facebook* (New York: Penguin, 2018).

Standage, 2013, *Ibid*.

Ito, Joichi and Jeff Howe, *Whiplash: How to Survive Our Faster Future* (New York and Boston: Grand Central Publishing, 2016).

Benkler, et al., 2020, *Ibid*.

Chapter 2.3

Alexander, Bryan, *The New Digital Storytelling: Creating Narratives with New Media*, 2nd Edition (New York: Praeger, 2017).

Csikszentmihalyi, 1990/2008, *Ibid*.

Chen, Jenova, "Flow in Games (And Everything Else)," *Communications of the ACM* Vol. 50/4 (March 2007); pp. 31-34.

Johnson, Samuel, *The Works of Samuel Johnson*, (London: Bohn,1854) p. 143 at:
books.google.com/books?id=6JEwAQAAMAAJ

Johnson, Steven, *Wonderland: How Play Made the Modern World* (New York: Riverhead, 2016).

Kay, Alan, 1986, *Ibid*.

Forrester, Jay, "Counterintuitive Behavior of Social Systems," *Sage Journals*, v. 16/2, February 1, 1971), pp. 61-76.

Rheingold, 2000, *Ibid*.

Meadows, Donella, 1997 *Ibid*.

Senge, Peter, *The Fifth Discipline*, (New York: Doubleday, 1990).

Licklider, J. C. R., 1960/2003, *Ibid*.

Haymes, Tom, 2020/2, *Ibid*.

Land, George and Beth Jarman, *Breakpoint and Beyond: Mastering the Future - Today* (New York: HarperCollins, 1993).

Brown, Stuart. *Play* (New York: Penguin, 2010).

Chapter 3.1

Haymes, Tom, 2020/2, *Ibid*.

Weinbren, Daniel, *The Open University: A History* (Manchester, Manchester University Press, 2017).

Chapter 3.2

Haymes, Thomas, "The STAC Model: Rethinking the Basic Functionality of Informal Learning Spaces," *Current Issues in Education* 21/3 (Special Issue), June 18, 2020 at: https://cie.asu.edu/ojs/index.php/cieatasu/article/view/1915

Brand, Stewart, 1994, *Ibid*.

Brown, Tim, "Design Thinking Defined" at: **designthinking.ideo.com**

Johannsson, Frans, *The Medici Effect* (Boston: Harvard Business School Press, 2004).

Molloy, Jonathan C. "Can Architects Make Us More Creative? Part II: Work Environments," *Architecture Daily* (May 10, 2013) at: **www.archdaily.com/367700/can-architecture-make-us-more-creative-part-ii-work-environments**

Burkeman, Oliver, "Steven Johnson: 'Eureka Moments' are Very, Very Rare," *The Guardian* (October 19, 2010) at: **www.theguardian.com/science/2010/oct/19/steven-johnson-good-ideas**

Johnson 2010, *Ibid*.

Morville, Peter, *Intertwingled: Information Changes Everything* (Ann Arbor, MI: Semantic Studios, 2014).

Chapter 3.3

Nelson, Theodore, *Geeks Bearing Gifts* (Sausalito, CA: Mindful Press, 2009).

Shlain, Leonard, *Art & Physics: Parallel Visions in Space, Time, and Light* (New York: HarperCollins, 1993).

Sagan, Carl, *The Demon-Haunted World: Science as a Candle in the Dark* (New York: Ballantine, 1996).

Bush, 1945/2003, *Ibid*.
Engelbart, 1962/2003, *Ibid*.

Nelson, Ted quoted in *"Lo and Behold"* (Film, 2016).

Nelson, 1974/2003, *Ibid.*

Adams, Douglas, *The More Than Complete Hitchhiker's Guide* (New York: Random House, 1989).

Rheingold, 2000, *Ibid.*

Engelbart, Douglas, 1992, "Dynamic Knowledge Repositories," at: **www.dougengelbart.org/content/view/190/163/**

Pendleton-Jullian and Brown, 2018, *Ibid.*

McNeely, Ian F. and Lisa Wolverton, *Reinventing Knowledge: From Alexandria to the Internet* (New York: W.W. Norton, 2008).

Douglas Engelbart Institute at: **www.dougengelbart.org**.

Chapter 4.0

Wozniak, Steve "Surprise Tribute to Ted Nelson," (2014) at: **youtu.be/gl0Wfs70rV4**

Pendleton-Jullian and Brown, 2018, *Ibid.*

Nelson, Ted, "Eulogy for Douglas Engelbart," (December 19, 2013) at: **youtu.be/yMjPqr1s-cg**

Brynolfsson, Erik and John McAfee, *Race Against the Machine,* (Cambridge, MA: MIT Press, 2011).

Brynolfsson, Erik and John McAfee, *The Second Machine Age* (Cambridge, MA: MIT Press, 2014).

Frey, Carl Benedikt and Michael A. Osborne, "The Future of Employment: How Susceptible are Jobs to Computerisation?" (Oxford Martin, 2013) at:
www.oecd-ilibrary.org/fr/employment/automation-skills-use-and-training_2e2f4eea-en

Nedelkoska, Ljubica and Glenda Quintini, "Automation, Skills Use and Training," OECD (2018) at:
www.oecd-ilibrary.org/employment/automation-skills-use-and-training_2e2f4eea-en

Kasparov, Gary, "The Chess Master and the Computer," *New York Review of Books* (February 11, 2010) at:
www.nybooks.com/articles/2010/02/11/the-chess-master-and-the-computer

Nelson, Ted "Ted Nelson struggles with uncomprehending radio interviewer" (1979) at: **youtu.be/RVU62CQTXFI**

About the Author

Tom Haymes is Innovator in Residence and Innovation Strategist for Arizona State University's ShapingEDU Project and consults with several other colleges and universities on Technology Assessment, Space Design, Professional Development, Digital Communications, Curriculum Design, and Digital Futuring. Over the course of 40 years of technology experience, he has been Technology Director for a college of 20,000 students, a teacher, and manager of interdisciplinary innovation teams ranging from fully digital platforms to campus design. Most notably, he led the college design team for the West Houston Institute, a SXSW learning space innovation award finalist project for a Texas community college. He has written widely on topics ranging from technology adoption to military history and is an award-winning photographer. His book *Learn at Your Own Risk* (ATBOSH, 2020), discusses how to apply digital thinking to teaching in a pandemic. His writing can be found at ideaspaces.net and shapingedu.asu.edu. He tweets at @ideaspacesnet.

www.ingramcontent.com/pod-product-compliance
Lightning Source LLC
Chambersburg PA
CBHW070045080526
44586CB00013B/922